DATA MANAGEMENT FOR OCCUPATIONAL HEALTH AND SAFETY

A User's Guide to Integrating Software

DATA MANAGEMENT FOR OCCUPATIONAL HEALTH AND SAFETY

A User's Guide to Integrating Software

Constance P. Wrench
Medical Development Systems, Inc.
Deerfield, Illinois

VNR VAN NOSTRAND REINHOLD
New York

Library of Congress Catalog Card Number 89-38337
ISBN 0-442-20610-0

Printed in the United States of America

Van Nostrand Reinhold
115 Fifth Avenue
New York, New York 10003

Van Nostrand Reinhold International Company Limited
11 New Fetter Lane
London EC4P 4EE, England

Van Nostrand Reinhold
480 La Trobe Street
Melbourne, Victoria 3000, Australia

Nelson Canada
1120 Birchmount Road
Scarborough, Ontario M1K 5G4, Canada

16 15 14 13 12 11 10 9 8 7 6 5 4 3 2 1

Library of Congress Cataloging-in-Publication Data
Wrench, Constance P., 1924–
Data management for occupational health and safety: a user's guide to integrating software/Constance P. Wrench.
p. cm.
Includes bibliographical references.
ISBN 0-442-20610-0
1. Industrial hygiene—Data processing. 2. Industrial safety—Data processing. I. Title.
RC963.3.W74 1989
613.6'2'028553—dc20 89-38337
CIP

CONTENTS

PREFACE

Purpose of the book. This is not a "how to" or "do-it-yourself" book. Its purpose is to guide health and safety professionals through the unfamiliar field of software design and development. An integrated occupational health and safety computer system is the objective, and the book directs the reader's attention to topics that require thoughtful deliberation.

The two themes of the book. There are two essential themes, the first being that development of a software application must be a team effort. The team is comprised of the computer software professionals and the occupational health and safety professionals. The second theme is recordkeeping, which may appear to be a frivolous subject, since everybody in health and safety professions keeps records. Getting information out of the records, however, is a cause of frequent frustration, which is why these professionals need to look critically at their recordkeeping practices from gathering data to providing information.

The project. An integrated occupational health and safety computer system is a comprehensive project that is costly because of the development time it demands. It is not a project for the amateur computer enthusiast because it requires sophisticated database design and technology, skilled programming, and software design experience. Whether company management decides to develop the system in-house, to hire a consulting firm to develop the system, or to purchase a ready-made system, the book is intended to explain enough about database management and software development so that users can participate knowledgeably in the project. At the same time, it is hoped that the developer of this specialized software can gain insight into the database requirements of the occupational health and safety professionals. For the purchase of a commercial package, the book should provide criteria for evaluation.

The intended audience. There are many books by computer experts on all aspects of systems software development. While I make no claim to be a systems authority, I hope I have provided a "systems appreciation" course for the health and safety users that should, in turn, help the systems analyst and other development personnel in system design, development, and implementation. This book assumes that computer professionals are experts at developing an applications system, and that the technical requirements of the system are in the capable hands of health and safety professionals. Users will gain insight into what the system can and cannot do for the user, and will learn how to organize computer data for useful retrieval.

Developers will gain insight into the data processing specifications necessary for health and safety, as well as the meaning of a shared system that integrates data from the different disciplines of occupational medicine, industrial hygiene, and industrial safety.

Every business has its own operating idiosyncrasies. No occupational health and safety system can be designed to cover all perceived needs, but looking at examples of data, and how to use them, helps one to visualize how one's own data can be similarly used. Therefore, the illustrations and figures may extend the readers' vision of what will best serve their needs.

Organization of the book. The book is organized in five chapters so that the user and developer together can:

1. Look at the system possibilities.
2. Determine what resources are needed for an integrated system.
3. Help the user to appreciate the developer's role.
4. Help the developer to appreciate the user's role.
5. Help both the user and the developer to reason through retrieval and report design so that the system can be of maximum benefit to the user.

I am aware that this is not the normal way in which developers look at systems; they have their own sequence of development phases that runs from a needs requirement and systems analysis through implementation and evaluation. The departure from this typical progression was deliberate, since the emphasis is on interpretation of software development for users. I would never presume to tell systems people how to do their work. In fact, the user is constantly referred to the developer for assistance.

Information systems. In some organizations, systems specialists are in departments known as management of information systems, and they are collectively referred to as MIS. Since other companies may have data processing designations, I decided to refer to this group generically as information services. This action should not be considered direspectful; it was consciously done to differentiate between the occupational health and safety professionals and the systems professionals. I have considerable respect for systems expertise.

The shaping of an author. Like the occupational health and safety professionals I meet, I am personally acquainted with the passion for hoarding data. When I worked in asbestos research, I was forever finding new places to store my valuable records. Each time I was overrun with records, I found more efficient ways to use the same space for storage, eventually spilling over to cabinets in the corridor. Everyone in research revered records for the riches we knew they held, and when a researcher died, we hoarded his or her records along with our own because there was always the tantalizing thought that we might use them some day. We were pack rats, overcome by an insatiable passion for data.

Occupational health and safety professionals, too, exhibit this data-hoarding syndrome. On one occasion, I was visiting a young industrial chemist who conducted all the stack emission tests for a company. He wanted to show me some records, and escorted me to a room two floors below his office which was a former office overcome by the addition of file cabinets that had forced the human occupants to vacate. There was barely enough room to open the drawers. I can only attest to the fact that two of the drawers were packed to capacity with emissions test records, but the way in which he located the records was to recall the date of a related occurrence and riffle patiently through papers in search of that time frame.

Every detail of this seemingly inconsequential experience is etched on my mind because the man was suffering from a severe cardiac condition that made his every step a labor. He died shortly thereafter, and I continue to wonder what will become of all that valuable information. Who will ever again find those data? What contributions to the health and safety of that company's employees and to the community were lost by his tragic end?

If data are so valuable that they need to be saved, it follows that they should be preserved so that they can be found by the saver or anyone else who follows. Paper files served our needs when there was no alternative, but now computer technology can work for our benefit. Computer files not only save space, they offer efficient storage and they can help us find the data we want. Twenty years passed between my research experiences and my visit to the industrial chemist, but my incredulity was heightened by a more recent experience. When I was in research, I used to contemplate the tens of thousands of dollars annually that it took to fund the researchers, technicians, equipment, supplies, animals, and laboratory space for each of the dozens of projects like ours. On occasion, projects were abandoned before papers were published and data were shelved, a deed that I found appalling when I calculated the number of tax dollars it took to do several years of work. As for the chemist and his company, if the loss control manager had investigated and comprehended what it took to fill that room full of records, he would surely have had to double the dose of his ulcer medication.

Why consider a computer system? Loss control management could be practiced in every occupational health and safety office I have been in. I would urge each health and safety professional to calculate on an annual basis the cost in time to produce every record in the office, the cost of the space the records occupy, the cost of the cabinets and shelves they fill, and the cost of the time spent in trying to find data when needed. An examination of this sort is an eye opener because it reveals that the cost is likely to equal or surpass the cost of the personnel hired to ensure company health and safety. If this lesson were taken to heart, controls would be instituted in a hurry and an immediate effort would be made to computerize.

However, computerization is only half of the answer to the problem of loss control in data management. The other half is an evaluation of recordkeeping, revision of wasteful practices, and definition of how the computer can overcome losses. Computers are said to have revolutionized information management, but this is true only if the data that go into them are revolutionized first.

The appendices. Some of the ways to evaluate data are discussed in this book, but it would take another book to explore ways to improve recordkeeping. The practice of coding data is so important to computerization, however, that I have added an appendix to provide some guidance on how to approach it. The appendix is written in the first person because the information describes personal experience. The five chapters of narrative are more formally written because they are a distillation of experience, training, and research.

The computer problems of the safety manager in a small business have been addressed separately from the rest of the book. My deepest regret is that there are no simple computer system solutions that can be offered in this very special situation. Nevertheless, I suggest that the safety manager read Appendix A first, and then return to the chapters to look for whatever ideas may be meaningful.

The bibliography. A user's guide is not intended to be a scholarly treatise. For this reason, I have avoided extensive references to computer and occupational health and safety literature that would have produced pages of citations. Instead, I chose pertinent quotations from papers and books that I felt were especially appropriate to the discussion. Most entries in the bibliography are annotated in the event that a reader would like to explore further what I found to be such interesting reading.

CONSTANCE P. WRENCH
Fredrick, Maryland

ACKNOWLEDGMENTS

Everyone likes to perform well, but writing a book is a humbling experience that forces you to acknowledge your many shortcomings. You quickly appreciate the genius of Dickens and Hemingway as you try to express yourself, and as you consider your topic, you appreciatively recall the people who shaped your experience. You realize how many others contributed to this work that carries your name alone. If all the authors of this book were listed on its cover, there would be no room for the title.

I should like first to acknowledge the special contributions of my husband, Dr. John W. Wrench, Jr., and my friend, colleague, and boss, Allan R. Bushey, without whose support I could not have completed this project. My husband's abiding confidence in me supported me in more ways than I can describe. Allan Bushey has carried a heavier work burden so that I might have the time and freedom to write

Dr. Richard Chamberlain, President of Medical Development Systems, Inc., urged me to "Go for it!" when the book was proposed, an encouragement rare in business circles that speaks eloquently for the kind of company I work for. At no time has anyone in my company ever suggested what I should write. What appears in these pages is entirely my own responsibility and in no manner should reflect upon Medical Development Systems, Inc.

Inasmuch as I am a better user than a computer software developer, I was very ably assisted on the systems side through manuscript review and comments by Irene A. Navickas of Navickas and Associates, Northbrook, Illinois, for whose systems knowledge I have great respect.

Dr. Neil Hawkins, an industrial hygienist with Dow Chemical Company, and Mr. James Unmack of the Los Angeles Cal/OSHA office helpfully reviewed the section on assessment strategies for workplace exposures.

Too numerous to list are persons who have influenced my views as co-workers, the clients I have worked with, and my excellent professors at Frederick Community College, who enabled me to see occupational health and safety in exciting new ways through data processing and systems analysis.

Each of the persons cited has enriched this book in measures great and small, but the responsibility for any errors or omissions I claim to be solely my own.

Chapter 1

MEDICAL, INDUSTRIAL HYGIENE, AND SAFETY DATA IN AN INTEGRATED COMPUTER SYSTEM

Computer systems can be integrated in several ways, but the system discussed in this book is an integration of occupational health and safety data management through a single database that tracks workplace exposures and employee health to company advantage. Through tracking of hazardous agents and potential human exposure, employee injury and illness and environmental impact can be controlled, thus reducing unnecessary costs and legal liabilities.

A system is designed in accordance with company policies and goals. Through the combined efforts of users, management, and a software developer, a system proposal is submitted to management for approval after careful evaluation of its cost-effectiveness to the company.

The common goal of governments, both in the United States and elsewhere, is to safeguard human health and the environment. The burden of this responsibility rests with employers everywhere and a valuable aid in attaining this goal is an integrated occupational health and safety computer system shared by occupational medicine, industrial hygiene, and safety professionals.

1.1 DEFINING AN INTEGRATED SYSTEM

Integration. *Integration* can be defined as unifying parts into a coordinated whole. A *system* is a group of diverse units that function together as an organized whole. In the world of computers, the term *integration* is used in more than one context. Integration has taken place in the following:

- Brands of equipment
- New releases
- Multiple-function machines
- Multiple-function software
- Local area networks

Integrated software for personal computing indicates several applications in one package, such as:

- Word processing
- Spreadsheets
- Database management
- Graphics

This capability, which became possible with the development of 16-bit and 32-bit microprocessors, permits the user to access a single package for a variety of uses for the data. Another example is desktop publishing, which merges word processing with graphics and uses a variety of printing capabilities.

Integration can also mean merging machinery, or system management. For instance, laser printers can do letter-quality printing in a variety of fonts and include graphics as well. When machine interfaces called *links* are used, a 3.5 inch diskette can exchange data between a laptop computer and a 5.25 inch floppy disk in a personal computer. A new release of software can update an earlier version by integration. *Emulators* can make one terminal look like the terminal of another machine. (Typical of these are emulators of IBM 3270 terminals or DEC VT100 terminals.) When the technical specifications of machines can be matched by emulator software, users can combine different products in an integrated system. The integration process to be discussed in this book is computerized data integration for specialized use in occupational safety and health. The principal modules of the occupational safety and health system are occupational medicine, industrial hygiene, and safety management, and there may be many associated smaller modules. Linkages bind modules together for sharing of data between disciplines (Figure 1.1). Some data, such as most employee medical test results, cannot be shared because of confidentiality, although there are exceptions. Biological monitoring results can be shared by the industrial hygienist and the company physician to determine the exposure that an employee has received of a hazardous substance.

One area shared by professionals in all three fields is a library of Material Safety Data Sheets (MSDSs) used for ready-reference purposes. This information should be available for access at all times.

Purpose of the Integrated System. The principal purpose of the integrated system is to monitor the health of the employee and the workplace environment. A secondary purpose may be to track emissions from the plant that contribute to environmental pollution of the surrounding community. This raises the question of how broad the scope of the system should be. The answer depends on the size of the company, the number of employees, the computing resources, and the responsibilities of the

INTEGRATED HEALTH AND SAFETY SOFTWARE

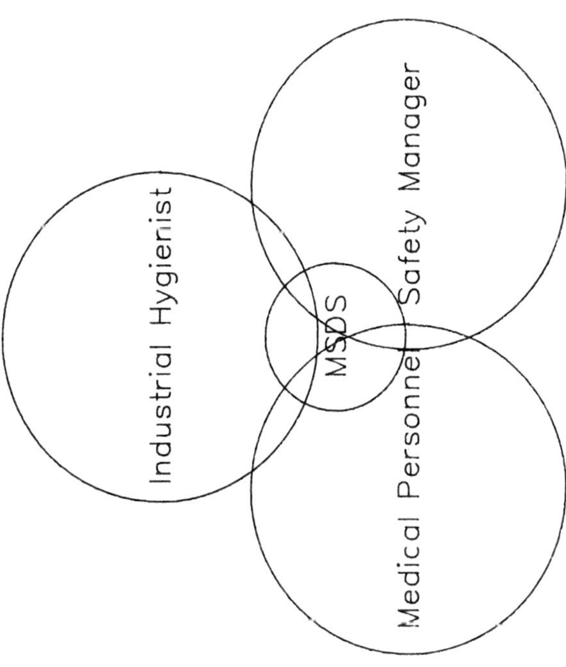

Fig. 1.1 Integrated occupational health and safety software.

principal departments. The occupational health and safety system is comprised of diverse data. These data are related through the employee and the job(s) assigned to the employee. Employee identification and the identification of hazardous physical and chemical agents are basic data links (keys) in the system. Other keys may be job assignments, processes, plants, departments, and locations. Medical test data for an employee, medical histories, workplace sampling data, and accident data together provide a picture of an employee's job-related exposures to hazardous conditions.

Necessity for an Integrated System. Occupational health and safety is an integral part of company operation, as are personnel administration, accounting, or manufacturing of the product. The company is responsible for those it employs. Samuels (1977) emphasizes that it is the company that bears malpractice responsibility, not the company physician, for instance. Therefore, it is incumbent upon the company to carefully record and preserve data pertinent to employee health and workplace conditions.

Convergence of law, ethics, and science in occupational health has been skillfully explored under the editorship of Lee and Rom (1982). Medical and industrial hygiene data are technical measurements derived from sciences such as anatomy, physiology, biochemistry, physics, and chemistry, to name a few. Once taken, these measurements become facts. Safety accidents are reported as a collection of investigative facts. Every measurement and every investigative fact recorded in the company records is subject to possible use for legal defense of the company in court. The government mandates certain responsibilities to the company for protection of the worker from occupational hazards and for protection of the public from plant emissions. Government mandates reflect modern ethical concepts of manufacturers' responsibilities.

The trend in occupational health and safety is a proactive approach that prevents occupational injury and illness. Potential health hazards are assessed, employee health risks are evaluated, and controls are developed to minimize workplace danger to the employee. Employees who participate fully in company preventive programs contribute to a working partnership for profitable production at reduced risk. Ideally, if occupational professionals pool their technical expertise and are supported well by management, absenteeism and health care costs decrease and production and profits increase. The point is that the real foundation of occupational safety and health in a company derives from all of the people involved, management and employees alike.

Planning a System. Developing occupational health and safety software is like building a house. Consider the style of the house, what rooms serve the purposes of the family members, and where the house will be located. This is the beginning. Consult an architect for discussion of the basic plans. The architect needs much more information, but these early ideas serve as a project beginning.

As an analogy to the development of occupational health and safety software, the consumer is the software user and the architect is a systems analyst. Software will be designed by the analyst only after basic ideas have been outlined by the users. The architect will ask many questions once an initial broad description of the project is given, and the same role will be played by the systems analyst.

The user must specify the basic requirements and scope of the application software. The systems analyst will look at the major parts of the system, reviewing every part independently and their working relationships with each other. The analyst will also do a requirements analysis by first examining present operations in the clinic, industrial hygiene, and safety, and combining this with a "wish list" for future procedures and work practices.

The Data. Data collection and maintenance is the basic function of the system. When the data are amassed and coordinated, they will be reviewed, analyzed, and reported for apparent relationships between employee and workplace. Management of the data is a separate function from reporting and analyzing the data. Both are part of the integrated system when the system design, software, and hardware permit.

Does an integrated occupational health and safety system mean that every activity under the purview of the medical staff, the industrial hygienist, and the safety manager will be included? The analyst will suggest limits to the system. Company constraints, such as development resources and cost, will automatically limit the system. The user should concentrate first on data that are required for employee and community right-to-know, data that must be used in order to identify changes in employee health connected with the workplace, or data that describe how well equipment and process controls are working. A natural tendency will be to incorporate every kind of record and report required of the professionals.

This may not be desirable from a systems point of view. One area of importance to the systems analyst will be the design of a database. Software applications are frequently designed using *the database approach*. The purpose is to unify and link all of the data required. Data integration and databases will be discussed in Chapter 2. Meanwhile, bear in mind that the present discussion about integrating a system does not include *data integration*, a term to be discussed in more depth in another section.

The Role of an Interface. A house totally isolated from the community would be unacceptable for family living. Typically, a house may have a walk connected to the front sidewalk or a driveway leading to the street. The family must buy groceries, clothing, and other merchandise. Family vehicles shuttle people to stores and back via the driveway to the rest of the community. These paths are interfaces that keep the family functioning. The occupational health and safety system must be constantly kept current (updated) by data from other departmental systems. Demographic data from personnel and job history information from payroll are good examples. Software interfaces establish connections to departmental databases and are developed in addition to the occupational health and safety software. However, the interface is not an integral part of the occupational health and safety system. It only serves as the pathway for sending required information into or out of the system.

1.2 ADVANTAGES OF AN INTEGRATED SYSTEM

Benefits of an Integrated System. Development of an integrated computer system will contribute to proactive occupational health and safety programs by weaving together the varied data of very different disciplines. Some benefits of an integrated occupational health and safety computer system are shown in Figure 1.2.

Administration. An effective, large database on a large computer system should be directed by a database administrator, or similar authority, who will coordinate system use and supervise maintenance. When the system is designed so that several groups of users share an integrated database, each group tends to input data in its own style, leading to data inconsistency. Employing a database administrator prevents unauthorized access and data inconsistency and removes the burden of day-to-day operations from the users. It may be desirable to appoint one person from within each discipline to be a local contact regarding database use or problems. The database administrator could be considered the general contractor in building the system, the person who oversees the job as it is being done. Where data are shared, consistency is a necessity. Different users will have different ways of saying the same thing. The database administrator must find terms that are acceptable to all users. The need for standardization will become apparent especially when searching for data. As an example, probably the most vexing problem for epidemiologists is data that are inconsistent in presentation and meaning. When data is preserved for many years, a significant effort must be made to maintain consistency.

BENEFITS OF AN INTEGRATED SYSTEM

Administration

Easier database administration

Large volume database

Easier data management

Reliability

Long-term data preservation

Data consistency

Sufficient data for analysis

Data for government compliance review

Data for comparison between plants

Retrieval and reports

Efficient production of ad hoc reports

Efficient production of standard reports

Organized for "right-to-know" requests

Organized for statistical evaluation

Cost-effectiveness

Single database cheaper to develop

Single database cheaper to maintain

Helps control health care costs

Fig. 1.2 The benefits of an integrated system.

Maintenance of the database should have a high priority in the company. From legal and economic viewpoints, reliable data may spell the difference between successful legal defense and paying court settlements or government fines for violations. Employees, their unions, and the public have a legal right to examine some company data on toxic substances; safety accident data may be reviewed by insurance companies. It is the responsibility of the users to furnish quality data to the system, and it is the responsibility of the company to preserve the data with care.

Integrating a system on microcomputers requires the perseverance and cooperation of several users during development and subsequent use. Where every user has a desk computer dedicated to his or her special tasks, integration is unlikely to occur. The importance of integration is that shared data are of interest to more than one user. The success of integration is measured by how freely the professionals share data with one another. The first requirement is to identify data of mutual interest. Once this is accomplished, a microcomputer database can be established. Mutual benefits can be achieved only through joint study, joint decisions, and close working relationships. Data can be shared by printed reports or by the passing of floppy diskettes. For a standalone microcomputer system data administration is carried out by the user, whereas in a multiuser system like a network or of microcomputers, there is an acting database administrator. Specific suggestions for building in data reliability, retrieval, and reporting on microcomputers will be discussed in later chapters.

In companies with fewer than 100 employees, a microcomputer system should have an internal hard disk with a minimum of 60 megabytes of storage. Now on the market are 160- megabyte hard disks, which are preferable. The anticipated 650-megabyte optical disks will support even more employee data for desktop occupational health and safety. Thus, no company, however small or large, need be without an an integrated occupational health and safety system because the hardware is available. The chief problem will be to find integrated software. Integrated software on mainframes is so expensive that smaller companies cannot afford it. The urgent need for microcomputer software will drive the occupational health and safety applications market.

Reliability. Ultimately, epidemiologists and toxicologists will review the data to follow exposure effects upon the employees. To reduce uncertainties in the data, industrial hygienists and medical personnel will include results of repetitive sampling and medical tests, leading to more accurate conclusions by the epidemiologists and toxicologists when they can be confident that the data they review are consistently reliable.

A database of numerous repetitive measurements is not a trivial accomplishment. Consider building a brick house from the ground up. When the bricklayer sets the first row of bricks, the pattern is cast for every row that follows. This is like a preplacement exam. The exams that follow will reference the beginning measurements. If there are missing or poorly aligned bricks in the wall, the stability of the wall will be in question. The same is true of the occupational health and safety system. If the foundation of the integrated system is reliable, repetitive data will engender confidence in future conclusions.

Sensitive measurements in screening and detection systems provide sound data. As detection equipment becomes more sophisticated, future measurements may be more sensitive than earlier ones. It is necessary that these changes be recorded in the documentation of an integrated system to avoid distorting analyses and conclusions. For valid statistical comparisons, the format of the data should be consistent.

System reliability is not only the product of computer hardware and software. It follows from system administration, reliability of the instruments and measurements obtained by the users, and user adherence to consistent work practices. A further requirement is the constant support of timely and thorough documentation, whether the system is small or large.

Retrieval and Reports. Reports are generated by the physician, the safety manager, and the industrial hygienist, each detailing information from his or her particular area. It is virtually impossible for managers reading reports of a diverse nature and presentation to obtain a clear picture of employee health and safety. An integrated system should unify the data and provide comprehensive information. The data used in the occupational health and safety system must be examined for shared areas that will benefit monitoring of the employee and workplace and will report results as clearly as possible. When a common cause for collecting data is achieved, a common method of reporting will also be found. This enhances the position of the occupational health and safety professionals with management. Management will welcome this.

Cost-Effectiveness. The strongest argument for integration of data is cost-effectiveness. However, it is almost impossible to pin down costs and benefits, except for a specific system and a specific company. A few measures to consider for a cost/benefit analysis will be given in Section 1.6.

Using a large and vital body of data covering medical test records, industrial hygiene sampling, and safety accident investigations, an integrated system will be less costly to maintain than several separate systems. Savings will occur in personnel, equipment, development, and maintenance.

1.3 SHARED DATA

A few general ways to share data will be discussed. Users will quickly envision others. The intent of this book, however, is not to describe every possible illustration of each topic, even if that were humanly possible. The intent is to stimulate the reader's thinking and to provide suggestions for implementation. As users plan their system, data of common interest can be identified, using a list similar that of Figure 1.3. In a company where industrial hygiene and safety are directed by a single person or department, there will be more sharing with the clinic than in a plant with three distinct departments. An integrated system can be advantageous in promoting the use of shared data.

Job Assignments. Job assignments involve all three areas. For preplacement exams, the safety manager will need to inform the clinic of the physical demands of the job and the physical and chemical exposures present. If a job safety (step) analysis is made, the safety department can be very specific about job demands. If this information is available on-line, the physician can easily determine the applicant's physical suitability for the position. If an employee has been on sick leave for

ON-LINE DATA SHARING

Data	Maintained By	Viewed By
Job Safety Analysis	Safety	Clinic and Industrial Hygiene
Equipment Changes	Safety	Industrial Hygiene
Controls Changes	Safety	Industrial Hygiene
Processing Changes	Safety	Industrial Hygiene
Medical Care Costs	Safety	Clinic
Disability Costs	Safety	Clinic
Training	Safety	Clinic
Injury Reports	Clinic	Safety
Biological Monitoring	Clinic	Industrial Hygiene
Employee Exposures	Industrial Hygiene	Clinic
MSDSs	Industrial Hygiene	Safety and Clinic

Fig. 1.3 Sharing data on-line.

illness or injury, the on-line systems's ability to review job demands will help the clinic determine whether to return the person to full duty. Where the clinic finds reason for concern, review of the job demands could lead to job rotation to shorten the length of employee exposures.

Even office and desk-bound jobs have their hazards. They can be physically stressful as a result of lighting, ergonomic, or other problems; they can be mentally stressful as well. This information can be useful to the clinic. Every office and department makes demands on the well-being of its employees. Most plants have no job step analyses because, however desirable they may be, the company lacks the resources to develop them. Yet, with on-line job analyses, the clinic is at an advantage in making decisions about work-related employee health.

When any changes are made in equipment, controls, or processes, safety and industrial hygiene personnel can update the system for exposure values, new equipment, process code numbers, and other relevant data. Medical personnel should be able to rely on an updated system without requesting information by phone or the plant mail system. Over time, the clinic may relate shifts in employee medical test data back to these changes.

Medical and Industrial Hygiene Data. There is a close relationship between medical and industrial hygiene data. The effect of hazardous exposure is determined by comparing medical test data with sampling results. Medical personnel should be able to reference the various kinds of employee exposures. The clinic should not attempt to interpret sampling measurement data, but exposures and permissible standards can be viewed. A given employee may show greater sensitivity than the average person. The hygienist can legally access certain records for biological monitoring, such as blood and urine tests. An alert industrial hygienist will note individual differences in employee susceptibility. There are no legal restraints on medical personnel viewing industrial hygiene datad although there may be company limitations. The important point is that a system contains data to be shared, and this sharing will improve the plant's occupational health and safety program.

Medical and Safety Data. A close relationship also exists between medical and safety data. Safety incident reporting and medical First Reports of Injury/Illness are examples. In some plants, First Reports are made out by medical personnel; in others, safety personnel perform this function. Data are then used for (OSHA) reporting requirements. Workers' Occupational Safety and Health Administration compensation data are of special interest to safety management, and injury or disability payments are of interest to medical personnel. Any contest or litigation over the degree of disability involving the company will be of special interest to both medical and safety personnel. When medical care costs, sick pay, and absenteeism become a company concern, safety management will share data with the clinic for review and recommend programs to reduce increasingly heavy company expenditures. Improved training programs may be one of the program's goals. Medical personnel may offer training as well as safety, increasing employee awareness and encouraging employee participation. Hopefully, this will lead to reduced health care costs. This type of data sharing is not on an on-line basis, but it should be considered in system planning. Which departmental data can be reported that will be mutually beneficial in keeping health costs reduced?

There are two types of data sharing. The type discussed above is sharing of on-line information. Recalling that information is communicated knowledge, the second type of information sharing can be in the form of reports between departments. Some forms of information sharing could be:

- Risk management and safety. Risk managers need data on incident costs for both personal and property losses. They also need information from the industrial hygienist about the potential severity of hazards present in the plant.
- Increasing incorporation of wellness activities into company health programs by occupational physicians. This may be done as part of the medical database or subordinated to the larger database. There can be programs to monitor blood pressure, achieve fitness, do aerobics, lose weight, or stop smoking. Medical personnel will want to know about employee progress in these areas through access to program records.

Data for General Use. MSDSs should be accessible not only to professionals in medicine, industrial hygiene, and safety, but also to plant personnel involved in purchasing, handling, and disposing of hazardous materials. The purchasing department may need equipment and materials specifications. The shipping and receiving department may need data for packaging, labeling, storing, handling, and shipping. The receipt of materials should prompt updating of the hazardous materials inventory and its storage location in the plant. Plant personnel should have access to handling and storage precautions; data on what to do about spills, leaks, and disposal; and emergency first aid procedures. Training personnel can refer to data on hazard ratings and toxicity. Clinic personnel need specific information on the materials being handled by employees.

MSDSs can be voluminous. Instead of having numerous copies of them strategically located throughout the plant, on-line access is more effective for general use. The MSDS additions and revisions can be made readily available on-line. Changes and revisions should be limited to one specified office or person.

A Word of Caution. Sharing data on-line does not automatically grant the privilege to both read from and write to the database. Sharing data is primarily intended for read-only use. Permission is granted to read only, write only, or both read from and write to the database. Reading from the database means that the user can only view data on the screen; writing to the database permits the user to add, change, or delete data from the database. A special need may arise for more than one department to enter data. An example of this is First Reports, where both safety and medical personnel need to contribute information on an incident. Access for both reading and writing can be controlled by passwords and other security features. Sufficient methods exist on larger systems for effective control. On microcomputer systems, decentralization of the database is, in itself, a deterrent to abuse of access for writing.

A multiuser system may present problems for users who attempt to write to the same variables at the same time. This is one of the issues that a database administrator can manage for everyone's comfort.

1.4 CENTRALIZATION VS. DECENTRALIZATION

It is helpful to understand centralization and decentralization of computer systems through examples. The growth of the Grow More Company, located in Chicago, will explain the concept. This company owns a minicomputer and has a small information services staff to operate the system and assist users. Users communicate with the central processing unit (CPU) by terminals connected through wiring in the building. All of the computer's power is concentrated in a single location. This is a *centralized system.*

Several years ago, the Grow More Company acquired the Dun Growin' Company of Boston. Dun Growin' also had a centralized system similar to that of the new parent company. Each of the two locations of the now expanded Grow More Company continued to operate its own system. Thus, expenses covered two computers, two staffs, and two sets of software. The effect was to decentralize computer operations for Grow More. The Boston group, accustomed to autonomy, strongly resisted combining the two systems into one, which is not unusual after acquisition. As time passed, however, the Chicago management noted significant differences in the data reported by Chicago and Boston.

Recently, Grow More purchased a St. Louis business called Still Growin', where each department had its own microcomputer system and the user departments administered control over their own data. In this company, computer operations were decentralized within the company. The users liked it this way. Their autonomy permitted them to decide for themselves what software and data they wanted to work with. Just before the acquisition, nevertheless, the management of the Still Growin' Company agreed to the need for shared data, and they wired their microsystems together into a local area network (LAN). A *network* is a series of interconnected computers. A *local area network* links computer equipment together in a loop within a localized geographical area. A local area network is an integrated system comprised of personal computers, file servers, print servers, and communications equipment.

After the purchase of the Still Growin' Company, Grow More's management decided that it was imperative that the computer systems and services be brought under control for both economic and administrative reasons. They bought a minicomputer for the St. Louis group, attaching it to the LAN. The decision was to have a *distributed system*. In this system, computer power and data would operate in two ways. Specific types of data would be collected and processed in local offices. Data of archival importance or for central decision support would be sent up to a mainframe purchased for the home office in Chicago. Boston was also linked by public communications lines to the Chicago office, completing the distributed system. This was now a distributed computer system networked to a central office. The distinguishing features of the distributed system are:

- Computing power in several locations
- Linkage through public utility communications lines
- A locus of centralized computing power

The outlying locations do not necessarily need to be linked to one another. Processing is allocated functionally across the interconnected computers. There is no shared memory by the computers, only message passing.

Which system is preferable for occupational safety and health? There is no single answer to this question. The decision about the ultimate design of the system is made after the systems analyst completes a requirements analysis and proposes designs suitable to the management structure, user needs, economics, database administration, and computer operations of the company. If the decision is in favor of a distributed system, the system design will partition the processes and data files, and allocate each to the appropriate computers. A single-location company with a minicomputer will probably have a centralized system. Where there are several locations, the company will have to consider centralization vs. decentralization. When users are accustomed to having their own desktop computing systems, they will likely be reluctant to give up their independence.

In addition to making decisions about the system, there is a related problem of centralizing or decentralizing the database. A *database* is a collection of different types of data related to a principal set of records. The database is designed to eliminate as much duplication of data (redundancy) as possible. In an integrated system, most of the database for occupational safety and health will be accessed through common fields such as employee ID or name, the employee's job code, or an agent name or ID. Databases are partitioned into sections, and the partitioned sections are allocated to groups of data. This avoids the problem of having users process the same data at the same time, an occurrence that is referred to as deadly embrace and that can lead to locking of the file so that it cannot be read from or written to. With a multiuser system which permits more than one user to use the system at a time, files are allocated by the operations software to individual machines and/or users. Clinic personnel can access medical data, industrial hygienists can access environmental data, and safety personnel can access safety data. In a system consisting of several microcomputers, it is easy to confine data to the special groups. In a larger system, access to grouped data may be by password or other security measures.

Systems personnel administer both machine operation and database control in a centralized system. All modifications to the database must be made through a central administrator, usually the database administrator (Figure 1.4a). This helps to keep the data organized for a common goal, even though the database is partitioned. It is also more economical to staff a centralized system.

The disadvantage is that administration is usually controlled entirely by systems personnel, giving users the feeling that they are not in control of their own interests. Occupational health and safety personnel are typically reluctant to relinquish control over their data. The problem can be partly overcome by having liaison personnel in each department work closely with systems personnel and each other. If this type of system is to be effective, it requires active communication between users and systems personnel. Other disadvantages to users are that systems personnel are not conversant with specific user problems and data, and there is often a slow response to users' needs because of the number of systems controlled by information services personnel. The slow response to user

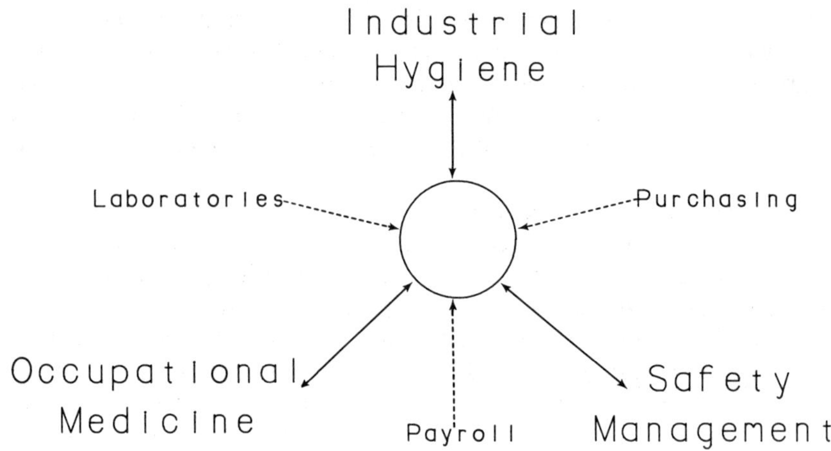

Fig. 1.4a A centralized database.

needs becomes an acute problem when users seek maintenance and repair on their systems or when they need additional development.

In a decentralized system, administrative control over the database is likely to be in the hands of the users (Figure 1.4b). The problem posed by this arrangement is that the database loses its integrated identity and may deteriorate into a collection of disconnected databases. An epidemiologist attempting to reconcile environmental and medical measurements is often frustrated by the incompatibility of data. What pleases the individual users may not benefit the organization as a whole.

In a distributed system, processing of the data is allocated to the individual machines in the system. The benefit of this system is that cooperation in achieving a common goal results from physical integration of the system's design. This is a more costly design because it requires more computer power, higher communication costs, and more systems personnel. Computer power resides in each location, and computer power is never cheap. These costs should be compared with the cost of expanding memory in a centralized system (also high). However, computer power increases processing speed, and on a centralized system users complain about being queued for processing. The centralized system loses speed in response time as the amount of user time or the number of users increases. A distributed system provides responsiveness and access while offering user independence.

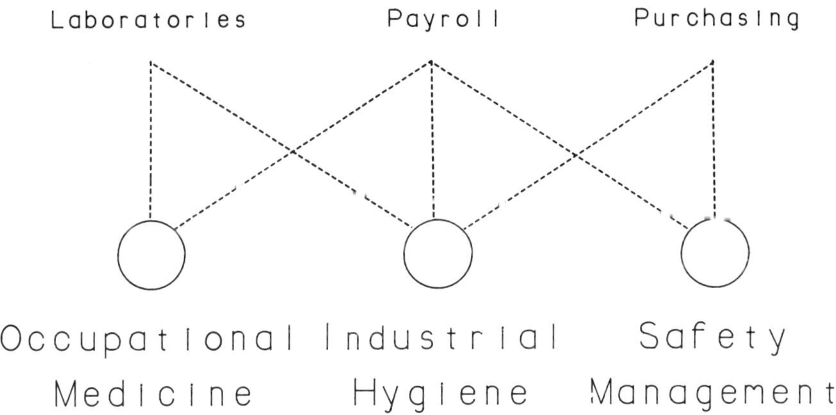

Fig. 1.4b A decentralized database.

In designing the system, the systems analyst must allow for growth potential. Can more users be accommodated and not diminish computer effectiveness? Can the database be modified as the dynamic occupational safety and health field evolves? What will be the future storage demands, and can the data be archived safely over a long period of time? What will growth require in terms of machinery and dedicated personnel? Can the system accommodate software changes, and how is this likely to affect the database?

Cost and effectiveness are difficult to measure in most occupational safety and health automation. This is not true of the overall design of the system. The costs of hardware, software, operations, and staffing can be defined. For example, one can compare the costs of using communications in a distributed system versus a centralized system with terminals wired directly to the computer. Time differences for accessing the files can be quantified by determining how long it takes to transfer a given amount of data. The most costly initial outlay for the distributed system will be for hardware and software. The expense of dedicating personnel to the tasks of data entry, maintenance, and administration obviously will be greater in the distributed system. The item that does not lend itself to quantification, however, is the personal commitment of users to participate fully in the ongoing tasks of database and system maintenance and administration. The comparison of distributed and centralized systems in this regard depends almost entirely on the people involved.

1.5 CORPORATE VS. LOCAL CONTROL

Control. Control is a normal function of management. In an occupational health and safety computer system, control raises the question of who provides system management, those who use the system or those who maintain it. For a small company there is little concern, because the entire responsibility for health and safety is usually assigned to one manager, either safety or medical, and control of the computer system will rest there. For large companies, the subject of system control is worth examining.

Perceptions. Where there are different departments for medical, safety, industrial hygiene, and information services, the question of who will direct the system is inevitable. Unfortunately, an answer to this question often gets impaled on the spears of departmental power struggles. *Power* is described as the degree of influence in a company, and *influence* is perceived. The perceptions of both users and information services need to be understood.

Users. If the system is to have meaning for the users, they want to exert control. The system will help them do a better job for the company, and they will be in the best position to determine what the company needs. They are the consultants, the professionals, to whom management looks for technical advice. The company is their client. No one in information systems will be sufficiently specialized to determine what data are required or how the data must be interpreted. Consequently, the inclination for users to decide what the system shall contain and how it shall be run is normal. Nevertheless, users readily admit to having no expertise in developing a system, putting hardware and software together, and maintaining them over time. They do not doubt the need for support from information services.

Information Services. This department also has its professionals, whose expertise is in putting systems together and maintaining them. They are a service corps dedicated to serving users (clients) of the systems. Their chief client is the company. Their responsibility is to make sure that company information, one of its valuable assets, is protected. Part of this protection consists of making certain that incompatibilities in hardware and software do not become an unnecessary company expense. Another part is ensuring that there is continuity in data collected and reports generated.

Where Perceptions Collide. There are two sets of professionals in the occupational safety and health system: the systems people and the occupational safety and health personnel. Each has a significant role to play, but their contributions often turn into a struggle over who gets the last word about the system, according to Markus and Niels Bjorn-Andersen (1987). Systems analysts should provide users with courses of action and data by which the users can make decisions in developing their systems. A question discussed in Section 1.4 about centralization or decentralization of the occupational safety and health computer system is a case in point. The proliferation of microcomputers in individual offices decentralizes control, and leaves the users in charge of their own work, which is why users prefer microcomputers; it leaves them in control of their own problems and solutions. In shared mainframes and minicomputers, the users lose control to the experts in information services.

Markus and Bjorn-Andersen make some astute observations. In the heyday of large computer systems in the early 1970s the responsibility of information services was to control acquisition and operation of data processing. By 1987, the emphasis on newer technologies such as processors of text, graphics, voice, and video had shifted their interests. Information services now provides its clients with maintenance, training, and ongoing support.

Users are increasingly computer literate, and there is so much applications software sold by knowledgeable vendors that users tend to dictate purchases. Recent developments in applications software permit users to generate their own software in some cases. Behind a user screen interface lies the capability for the machine to generate programs. This eliminates the user's need to rely on information services programmers. Vendors sell the package with user training, which externalizes the entire process of developing software for clients. The availability of vendor services shifts user interest in getting assistance with their systems from information services experts within the company to the outside. No longer is it inevitable that information services will exert primary control over design and compatibility of hardware and software within the company; users assume control and take their business elsewhere.

An interesting problem arises when the user purchases a system that evidently meets all data processing requirements and is given a starting boost by the vendor. If the documentation is poor and the user has no idea what to do next, the vendor may have a telephone hotline. However, it is more likely that information services will be called to the rescue. Information services then requires time and resources to become familiar with the new system so that they can continue to service the user, who should have proceeded under their direction in the first place.

Common Goal. It is imperative that all departments work together to achieve a common goal of uninterrupted information flow throughout the company in an effective and useful manner. Whether the system is intended for use by accounting, payroll, personnel, or occupational safety and health personnel, it belongs to the company, not the users. If information services has been charged with oversight of computer use in the company, this department should bear in mind its service function and assist users in acquiring and maintaining the best hardware and software available within the company's capacity. The reader should not infer that this charge places power and control in the hands of information services. Information services must support the users with computer expertise.

It is the responsibility of the user to help decide what kind of system is compatible with the intended use and to use company resources conservatively. The users must play a very large role during the development phase of systems analysis and design. The benefit is that the user will not only acquire a system geared to the user's needs, but will also become familiar with the system's strengths and limitations. This type of education is invaluable for the future, when maintenance is called for and modifications are desired. It is good to know what information services has to offer, and it is comforting to know that they are there to respond.

Systems Analyst. The key person to balance the equation between users and information services is the systems analyst. The analyst's job is to keep the user happy while protecting information services from excessive burdens on their limited resources. Programmers may never know whether the user is content, and they certainly do not care to hear that the user has a gripe. Yet, it will be the programmer that the analyst turns to in solving a user's problem. With considerable tact and good humor the analyst will communicate with the two camps, making the user happy with the service received and the programmer delighted with rendering a service. The systems analyst is thus a customer relations specialist.

Future Benefits. The small company with no information services department is at a disadvantage. Occupational safety and health personnel must rely on vendors to assist them in solving their hardware and software problems, and the service is not consistently good. Taking advantage of newer technologies is not always possible, given the need to modify the system without disrupting the existing database. The large company can forge an ongoing partnership between occupational safety and health and information services that may promote an evolving data processing system. Medical practices change, industrial hygiene practices change, and safety engineering changes. These professionals will insist that their systems keep up with advances in their fields.

Microcomputers. Professionals in occupational safety and health now depend heavily on the use of microcomputers. These technological tools had an extraordinary impact on business and society in the 1980s. By the twenty-first century, we may be able to quantify, in restrospect, the degree of impact. What we can see at present is that microcomputers inevitably alter management and recordkeeping. The past expensive, large mainframes cost hundreds of thousands of dollars. Operation had to be carried out by computer specialists. Today, by contrast, bookkeepers, managers, and clerical help trained in microcomputer operation work independently. Users have computer power at their fingertips. Once a user becomes familiar with the advantages of microcomputers, new accomplishments are conceived and the user quickly envisions new spin-off applications.

1.6 COST/BENEFIT ANALYSIS

Cost/Benefit Ratio. The two problems with trying to suggest how to do a cost/benefit analysis on a computer system for occupational safety and health are that:

- Every system will be unique, although it will share features of other systems.
- There are some features for which there may be no conclusive way to determine the price.

What can be done is to suggest items to be considered for inclusion in the analysis.

Using some monetary means for comparison, a cost/benefit ratio can be derived to help users convince management to invest in an integrated occupational safety and health computer system. The two parts of this ratio are the costs of the system and the cost of the benefits derived from the system.

Tangible (Direct) Costs. Costs will occur in producing and operating the system. The initial costs will be like those shown in Figure 1.6a. Afterward, operating costs begin. How many personnel will be required to maintain the functioning system? If it is a mainframe or minicomputer, trained personnel will be required for many tasks, such as operating the machinery, loading software, backing up data, and recovering data. Backup procedures are vital daily tasks that protect user data. With a variety of users, a database administrator may maintain the data dictionary and develop and enforce user policies. In a small system, there may be little need for extra help other than occasional hardware repair, which

DIRECT COSTS OF A COMPUTER SYSTEM

Needs analysis

Requirements analysis

Systems analysis and design

Programming and software development

Additional software (e.g., DBMS package)

Hardware

Installation

Training

System implementation

Fig. 1.6a The direct costs of a computer system.

can be accomplished by the nearest vendor. For a large company, staffing can vary over several levels, from professionals to technical operators to data entry personnel.

Where the company does not already own the database management system (DBMS), this will be an initial cost but one can also be leased. Maintenance must be done on the DBMS from time to time, which is a cost to be factored into system maintenance.

If there is a central computer, will information services charge the user departments for processing (CPU) time? If this is the case, estimates will be required to determine the frequency and extent to which they will use the system.

One feature of operation invariably gets overlooked. Who will continue to document the system, once installed, to make certain that every change is recorded? This could be a function of the database administrator on a large system, but where microcomputers are used in several different departments, the user must delegate the responsibility to one person.

Include items like depreciation on the hardware, part-time employees'salaries, and fringe benefits for all employees. Hardware will include not only the computer(s) but also cables, terminals, printers, disk packs, and communications equipment. There will be supplies like paper, diskettes for microcomputers, magnetic tape, and ribbons for printers. What furniture, like desks, tables, chairs, and storage cabinets, will be required?

Intangible (Indirect) Costs. Intangible costs are also included in the cost/benefit analysis, although they are not readily associated with the purchase. Will there be a need for new room space? Larger computers require air conditioning. A requirement for added security should be considered, and the protection depends on the individual situation. One security measure should be adequate insurance for the computer system. What increases will be needed to supply the area with janitorial, maintenance, or other support services?

Tangible (Direct) Benefits. When doing a cost/benefit analysis for occupational safety and health, it is very difficult to describe accurately the monetary value of a computerized system. Chief objectives for the entire occupational safety and health field still need to be defined inasmuch as the field continues to evolve. In time, there will be established work practices and criteria for making judgments about effectiveness. Benefits of the new system may not be measurable with certainty in the immediate future.

Examples of important present additions are workman's compensation payments, disability insurance payments, wellness programs, and off-the-job safety programs. The primary characteristic of a working computer system is that the more acquainted and comfortable the user becomes with system, the more uses can be added for greater productivity and data handling. Characterizing this monetarily will be a challenge for both users and systems analysts together. Figure 1.6b lists some tangible (direct) benefits that result from a system.

Intangible (Indirect) Benefits. Employees and their union representatives who see the commitment of management to a comprehensive occupational safety and health program will respond with improved morale and interest because they see the program as being in their best interests.

Kinds of Costs. There are other ways of identifying costs. Initial costs are one-time charges. These will be chiefly for design, development, and purchase of the system. They will cover the requirements analysis, design of the system, programming or purchasing the software, and purchase of the hardware and DBMS.

Front-end costs will include the implementation and testing of the system. There may also be one-time charges, such as special user requests for training or copies of documentation in addition to those normally supplied. If implementation is phased, training costs could be multiplied. An interesting question to be answered about implementation concerns incorporating data in company files prior to computerization.

DIRECT BENEFITS OF A COMPUTER SYSTEM

Personnel time saving

Improved services

Faster access to data

Faster data retrieval

Reduction of errors

Reduction of lost data

Reduction in record storage space

Potential for expansion

Standardization of data

Improved reporting resources

Easier and faster employee notifications

Fig. 1.6b The direct benefits of a computer system.

The amount of data entry required in most cases would require provision for a data entry person dedicated to the task for a period of time that is proportional to the amount of data it seems best to enter. This would be an ongoing cost for a specified period.

Training costs need to be spelled out in detail since more than one kind of training will be required (e.g., for operating personnel, management, professionals, and data entry operators). The number of hours or days of training for each must be calculated. Training on-site is usually cheaper than sending everyone to the developer's location.

Documentation should come with the system at the time of implementation. The cost of updating documentation over a period of time needs to be specified in the original contract.

When systems are implemented, there are often good reasons to run parallel systems. Continue the old system while testing the new system until the user feels confident that data will not be corrupted or lost. The old system may consist of paper files or a simple computer file for limited use. Specify a cutoff time for running parallel systems, and calculate personnel and other costs for each system in this period of simultaneous operation.

Development Costs. The developer should be able to furnish a time schedule for development. There will be costs not only for the developer but also for the users. The amount of time required for the user departments to develop specifications and data requirements, and to spend time conferring with the developer, should be estimated.

The Cost of Change. Computers invariably dictate change. Changes affect people, work practices, and, inescapably, the company treasury. Will the new computer system require new personnel or retrained current personnel? In what way will the computer system affect the work practices of medical, industrial hygiene, and safety personnel? In the early period of use, the system will make demands on their time and energy instead of reducing their effort. A typical area of impact is in the design of forms, which have to be revised to be made compatible with the computer data input. Next, everyone has to be shown how to use the new forms and must take the time to get used to them.

1.7 CORPORATE ADVANTAGES OF AN OCCUPATIONAL HEALTH AND SAFETY SYSTEM

Company Assets. Protection of company assets is a high-priority item on the company's agenda. An occupational safety and health program makes a significant contribution in this regard. Among these assets are:

- The employee work force, the company's greatest asset. From the employees, the company expects high productivity, low absenteeism, and a minimal accident rate.
- Materials and processes for production, in which the company has a very large investment. This includes raw materials, equipment, products, and process intermediates, some of which may be injurious to employee health.

Protecting the Assets. An occupational safety and health information system can assist in a variety of ways:

- Updated records keep staff and management proactive instead of reactive with respect to changing conditions in the plant.
- Computerized data on employee exposures will help to identify early changes or trends in environmental conditions.
- Data tracking of process and equipment changes protects both employees and capital investments.

- Spotting of trends in the data permits early correction of equipment and processes, minimizing production costs.
- Underlying causes of accidents can be detected through detailed analyses of accident data, reducing personal losses, property losses, and compensation payments.
- Hazardous materials tracking controls storage and handling dangers.
- On-line MSDSs offer immediate references for handling, storage, and accident control of hazardous agents.

The company has an accounting department that daily monitors the relationship between receipts and expenditures, calling the balance cash flow. When the cash flow decreases by any amount, a search is instituted to find the cause, corrections are implemented, and an improved balance results. If the same effort were assiduously applied to occupational safety and health, not only would company assets be protected, but financial improvements would be detected.

In financial matters, management must identify problem areas at the earliest possible stage, hopefully as early as the planning stage. Immediate countermeasures are taken to eliminate imbalances. In the accounting department, the primary tool for keeping tabs on the cash flow is a computer system that speedily and accurately handles large volumes of data. The computer not only handles receipts and expenditures, but also links to payroll data, disability payments, and other cost areas to integrate critical information from other departments.

The same should be true for occupational safety and health. There are important data that can link between the clinic, the industrial hygiene department, and the safety department, showing the company a bottom line on their agent levels and employee health, identifying and evaluating the anticipated stresses on employees, checking further damage, and making corrections early before costly incidents occur. Productivity in occupational safety and health operations like accounting needs computerization, for the same basic reasons, handling volumes of data efficiently for quick retrieval, review, and decision making.

Competitive Edge. Ultimately, faster identification of problems helps to give the company an edge over its competitors. In an excellent speech by Dr. Lloyd Tepper (1987), the strategic value of occupational safety and health to a company is examined. Dr. Tepper shows the relationship between world economic conditions and the competitive edge given to a company by its careful attention to occupational safety and health. He draws a picture of the physician's role in the company and asks whether the physician sees himself in relation to his peers and profession, or whether he uses his professional expertise to solve the problems that the company faces. The occupational safety and health effort has to be made a vital element in the development of successful business strategies to meet market competition. Total integration of occupational safety and health into the company's management and organization is needed to accomplish this goal. If occupational safety and health is a paper program only, nothing happens; if no money is invested, nothing happens.

American business is aware of the competition that Japan has given the U.S. auto industry. The Japanese government joined forces with its auto makers in a united front. The companies needed equipment to increase productivity and their government made that possible. In this case, the government recognized that the overall economy would benefit by giving productivity a boost even though it required monetary outlays. Business benefited, and so did the Japanese economy.

In an analogous situation, buying equipment to make it possible for the occupational safety and health departments to best serve the company will ultimately accrue to the economic benefit of the company. It is necessary to build a workplace compatible with good employee health or face the probable economic effects of absenteeism, low morale, and government intrusion.

Can the Company Afford this System? The cost of *not* computerizing will be felt when records are needed for situations like those shown in Figure 1.7a.

For the company looking at a decision that will cost thousands or tens of thousands of dollars, the bottom line is a critical factor. If a cost/benefit analysis shows a positive benefit, the only question may be where to find the cash to pay for an integrated occupational safety and health system. In some cases, the need for information is so important to the company that a system will be bought relative to the company's ability to pay. Although most companies like to keep abreast of technology, they never find this a sufficient reason to invest in an expensive computer system. Yet, it is the constant improvement in computer technology that helps improve business management.

Without question, management will benefit from a system that reduces the response time to information requests. How frequently is information sought from departments related to occupational safety and health? How long does it take to search for the answers? When information comes from managers of medical industrial hygiene, and safety, are the results integrated, or must management make inferences about the general meaning of these segmented reports? A welcome feature of an integrated system is that it can merge data from two or more departments, thereby

USEFULNESS OF COMPUTER RECORDS

Defense against litigation
Verification in disability and compensation cases
Response to employee and community complaints
Public relations information
Protection against regulatory violations or noncompliance

Fig. 1.7a The usefulness of computer records.

relating data from different sources in a single summary. When management notes trends that need attention and correction, changes can be made. Separate summaries of data from the clinic, industrial hygiene, and safety make it difficult to detect early trends. The majority of U.S. companies with occupational safety and health programs do not merge data from several departments. The principal reason is the lack of resources to do the job. Manually, one person will be kept busy making these summaries, but on an integrated computer system, the reports can be periodic, comprehensive, and meaningful.

Biting the Bullet. Management, information services, and users will each have a different view of the occupational health and safety system (Figure 1.7b). The large company considering an outlay of $150,000 to $200,000 to automate occupational health and safety, or the small company weighing the expenditure of $25,000, will cringe at the cost of a computer system because this expenditure is considered a departmental outlay, not an investment to protect company investments and assets. Why is a tracking system needed at all? This question is best answered by asking another question. Why automate to keep track of accounts receivable, accounts payable, sales, payroll, or any other normal financial operation? To track human assets just as painstakingly as dollar assets! To accomplish this objective, the occupational safety and health system must be viewed as an integral part of company operations.

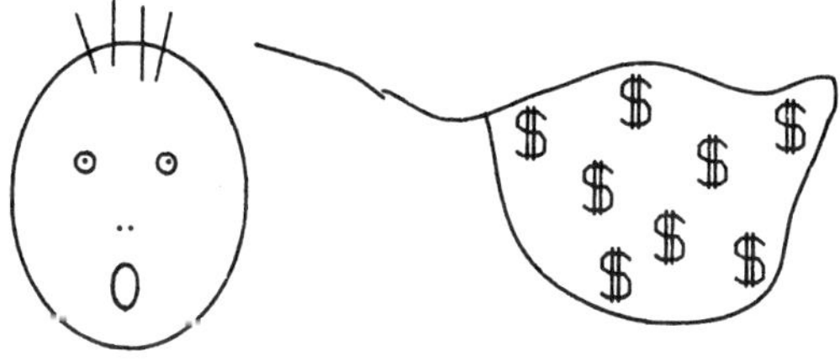

BOARD OF DIRECTORS

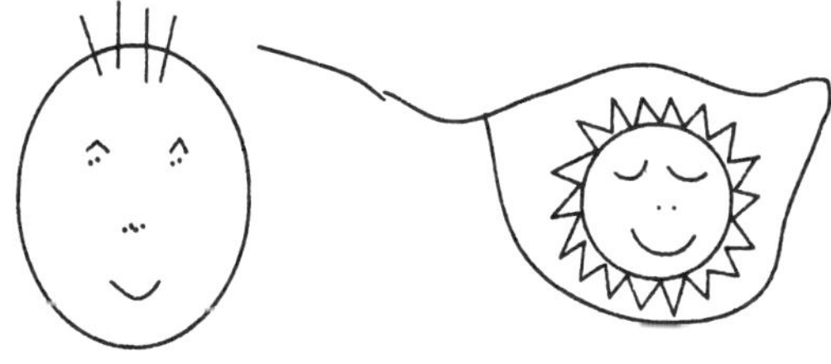

HEALTH AND SAFETY USERS

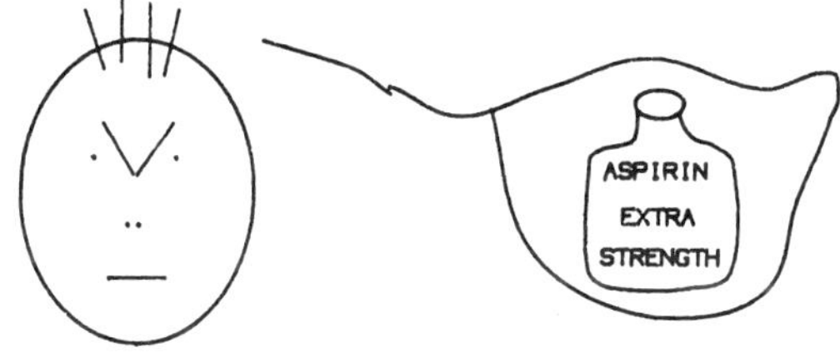

INFORMATION SERVICES

Fig. 1.7b Several viewpoints invoked by the anticipation of a computer system.

1.8 RESPONDING TO GOVERNMENT REQUIREMENTS

Government Expectations. What the government expects of industry is fourfold, as shown in Figure 1.8.

Occupational illness and injury is a subject as old as the human race, but only in contemporary history has it been given important recognition. We need data on its location, its extent, and its causes to guide us toward

GOVERNMENT EXPECTATIONS OF INDUSTRY

Protect employees from conditions that cause injury or ill-health

Keep employees informed of the potential for harmful exposure

Train employees to guard their own health and safety

Provide employees with necessary controls to protect against hazardous exposure

Protect the community from environmental exposure to industrial hazards

Keep the government informed of accidents and harmful exposures

Keep government informed of work-related employee illness or injury

Fig. 1.8 Government expectations of industry.

acceptable improvement measures. As the government collects data, national statistics on occupational injury and illness are compiled that make it possible to compare illness and injury rates among industries. There is a critical national and international need for this type of information. The data collected nationally by the Bureau of Labor Statistics at present are limited; more comprehensive industry-by-industry data are needed. Occupational disease is often difficult to trace to its origins because medical statisticians lack sufficient data for identifying the etiology of disease arising in the workplace.

Legislative Impact. The principal legislation affecting manufacturing was the Occupational Safety and Health Act of 1970. Requirements of this act emphasized monitoring employee exposures, monitoring employee health, training employees, and protecting employees. In response, corporate budgets were altered to add physicians, nurses, industrial hygienists, and safety specialists. Recordkeeping became a formidable task. Supplementary legislation on environmental issues increased the burden.

Workman's compensation legislation by states traumatizes record-keepers. Small wonder that employers feel so put upon. In addition to their worries over competition in the marketplace, they find the government requiring records and reports that can only be prepared at the expenditure of money and time.

Added to the impact of legislation on occupational safety and health are additional pressures on behalf of employees, such as union representation and litigation by employees and their families. Judicial interpretations of the law have broadened the obligation of the employer to provide safe working conditions. Company budgets will never be the same. Plants and machinery are redesigned, often at a cost of tens of thousands of dollars. Expensive testing is done on both employees and work sites. Training and retraining take place for off-the-job as well as on-the-job reasons. Health improvement and education programs abound. Despite all the burdens, however, the additional expense for the employer has been shown to be a worthwhile investment.

Ripple Effect. A ripple effect has resulted from the legislative and legal pressures.

- Employees are demanding a piece of the action. Where employers formerly had little to say about their job welfare, they are now vocal and are fully supported by unions and courts. Workplace conditions now provide more chips at the contract bargaining table.
- Employee health has become an important factor in the decision-making process. Will the machinery being purchased be safe? Will the process being evaluated be safe? If health benefits cost a certain amount, will the budget permit the addition of only two new employees rather than three?
- Hiring practices have changed. Employee selection criteria are carefully designed. A company can no longer simply hire a body to fill a slot. Job requirements are drawn with consideration for the physical and mental demands of each task. The more physically fit applicants are less likely to need health care. Healthy lifestyles promise less carelessness, fewer accidents, and less disease. Careful placement promises safer performance, less absenteeism, and more productivity.
- Sharper management skills and greater knowledge are required. One objective is to identify personal problems in the early stages and know how to make referrals. Recognize the potential for accidents

even before a near miss occurs. Provide leadership in health education and maintenance. Decrease workplace stress through education and training.
- Support services have expanded. Larger companies have built exercise and health education facilities. Smaller companies contract for these services with a local spa or YMCA. Company cafeterias promote better nutrition.

The ripple effect extends beyond the workplace. The employee takes home increased awareness of safety and good health. The precepts of the workplace become family practices, with lowered cholesterol and lowered sodium intake and more exercise, for instance. This emphasis on better health is subsequently reflected in reduced health benefits for the employee's family.

Typically, management initially responds to regulatory statutes for control of hazardous exposure in the workplace by grumbling about the cost of legislative interference in business. Legislative initiative has focused management's attention on the importance of the employees in achieving organizational goals and objectives. It is unfortunate that the government has to force some businesses to keep employee safety in view. What else can employees, unions, or the government do but throw a cloak of protection over employee health and safety? This sets the stage for litigation and government penalty.

The Reality. Protection of employees and the public from hazardous exposure should not be an *adjunct* of business operations requiring staff attention. It must be embodied in the business fabric itself, with direction from the chief executive officer and all top management personnel.

The government would operate at less cost if regulatory measures did not exist. But regulations do exist, and standards and enforcement follow. Employers are irritated by the need to use their resources to meet government standards, and they become irate over enforcement penalties when standards are not met. In fact, it is the threat of big legal liabilities that has promoted the growth of occupational safety and health programs, not the fact that it is good business to protect employees from exposures before the government gets in the door.

Good Business. Industrial organizations are not always at cross purposes with government regulation. There are some commendable examples of cooperation. The Celanese Corporation and the American Petroleum Institute are outstanding in this regard. Celanese (1983) supported OSHA's development of the Hazard Communication Standard with the following statement:

> As a concerned and responsible employer and producer of chemicals, we desire that our own employees and those of our customers be provided with accurate information on the hazards of the chemicals they handle and are instructed in proper work practices to minimize the risk from those hazards. We thus support OSHA's intention to develop a sound standard to support those aims.

At the same time, the American Petroleum Institute (1983) is quoted as saying: "API has consistently supported the concept that each employer should have a written program to detail how workplace hazards are identified, evaluated, and communicated to employees."

One of the costs of doing business in a democratic society is the need to protect society. The democratic process makes this clear when elected representatives of the people mandate protection of employees and the public from health hazards by creating agencies like OSHA and the Environmental Protection Agency (EPA). The power of standards enforcement comes with the mandate. When employers ignore or circumvent the standards, costly penalties will surely result. The prudent business will conduct an occupational safety and health program in the spirit of normal business procedures rather than as a program of defense against the government. The company should expect to control its own internal conditions, and not have to be controlled externally by regulatory actions.

An astute employee observed that "the competitor ain't our worry. Our lives is what we're worried about" (Weiner 1980). If the employer protects the worker, the worker is then able to concentrate on the business goal of making a competitive product. A typical concern of employers is that too much employee knowledge creates more problems because the employee or his union will use the data to obtain workmen's compensation, an expense item that will be added to the company's ledgers. However, says the Chemical Manufacturer's Association (1983), "There has been no serious dispute of the proposition that well-structured programs for informing employees of the hazards of the workplace are beneficial and cost-effective."

More Views of Government. Neither business nor government can lay claim to sole ownership of white hats. The government agencies have failed the private sector in some respects. Standards for critically hazardous materials need improvement in some cases and have not been developed in others. Standards are rigidly enforced in one place and ignored in others. Some standards are poorly defined and difficult for specialists to interpret. It is impossible at times to get answers from agencies. The government deliberates at a snail's pace.

The courts are equally frustrating. Legal proceedings take time and money. A company can spend a lot of money defending itself against charges by unions, employees, and environmentalists, to say nothing of staff time.

Additionally, the public gets frustrated with courts, business, and government agencies. John Q. Public has little patience with any of those involved. The media report with an accusatory finger toward business, often making it difficult for the public to understand the complex issues involved. The common response of the public is to applaud the government for leveling penalties for infractions, even though this may not be a fair assessment of the actual situation. This is not to say that the media intend to be unfair; reporters are simply not technically adept in the fields of occupational safety and health, nor are they specialized enough to understand the issues involved. The result is that the public usually gets an oversimplified and abbreviated report.

Recordkeeping. Included in legislative mandates are requirements for records. While some data are directly specified, most of the recordkeeping is implied. The length of time for keeping exposure records has been set by OSHA to cover several decades in many cases, not several years. Measurements made for comparison with exposure standards and medical test results are examples.

Large companies have tried to develop their own computer systems for occupational safety and health, often at heroic expense. Some are for mainframes, others for minicomputers. Several computer companies have offered time-sharing arrangements. Integrated software for microcomputers has been attempted. No unequivocal successes have been achieved to date. Two big reasons for poor success are lack of supporting computer technology and lack of long-term development money. Tools now available will contribute to success, but we will only be able to judge the degree of success retrospectively. There is a desperate need for commercially available software that is portable across hardware and operating systems, but the government wants the records kept now.

OSHA requires that occupational safety and health program specifications be written. The information that will support programs will be a bank of data in carefully maintained records. The volume of data will vary according to the number of employees, the number of medical and industrial hygiene tests, and the number of accident records.

An Automated System. If there is evidence in the plant that hazardous exposure is harming an employee, the employer must waste no time in correcting the situation and must provide the employee with adequate health care and education to prevent additional damage. One benefit of a comprehensive occupational safety and health system is that better detection of possible harm to the employee will occur before health impairment. The objective is to detect impairment in the earliest stages wherever possible. An automated occupational safety and health system will contribute to this timeliness. Company investment in a system that tracks employees and agents reduces expensive compensation and disability payments, to say nothing of employee physical and mental distress.

A significant resource of the occupational safety and health system in the event of litigation is a computer bank of information. The best scientific evidence of hazard potential can often be debated. When a company fully records its exposure and accident data, it is fortified in court to argue its efforts at environmental control.

International Occupational Safety and Health. Before leaving the subject of government regulation, acknowledgment should be made of employer responsibility for safety and health in other countries. The International Labor Organization (ILO) is a branch of the United Nations that develops international treaties and sets standards for labor, including occupational safety and health. Its membership comprises representatives from government, management, and labor (unions). Member governments are required to assess production and control materials of single and multiple substances that may affect safety and health.

At Convention No. 161 on Occupational Health Services (OHS) Part I, Article 8 recommends that "employers, workers and their representatives shall cooperate and participate in the implementation of the organisational and other measures relating to occupational health services on an equitable basis" (ILO 1985, 3). Article 4 at Convention No. 155 on Occupational Safety and Health and the Working Environment says that the employer's representives and workers shall have a policy "to prevent accidents and injury to health arising out of, linked with or occurring in the course of work, by minimising, so far as is reasonably practicable, the causes of hazards inherent in the working environment" (ILO 1981, 2).

In Sweden (Kleinman 1984) this is accomplished through safety committees in plants with fifty or more workers. The committee includes representatives of both management and employees. The Swedes decentralize authority for occupational safety and health by giving employees the authority and knowledge to help implement programs. Final authority on financial support is left to the employer, but employees have the right to stop work if a situation is considered hazardous.

Occupational safety and health is not solely an American concept. It is part of a worldwide recognition of the importance of health protection and the responsibility of business everywhere to support employee and public environmental protection. The United States is a member of the ILO. Sixteen conventions have written twenty-three recommendations up to 1985. Many of these are concerned with hazardous substances. Member nations frequently use the recommendations to construct their own laws, even though the ILO guidelines are not binding. European nations like Great Britain, Germany, and Italy are especially active in occupational health and safety, and American professional organizations keep in touch. There is much to be learned from these exchanges beyond our shores.

1.9 TRACKING HAZARDOUS CONDITIONS IN THE WORKPLACE

Why Use Computers? How can the computer help track hazardous conditions in the workplace? It is an adaptable tool that helps the user manage large volumes of records in the face of constantly changing conditions and dynamic growth.

Computers are used to automate a plant in many ways, as shown in Figure 1.9a. Each of these examples is a special case of computer applications representing only a few of the technological developments now available that offer improved control over events.

An employee is an expert at his or her job, but well-planned use of a computer can decrease the repetitive tasks that detract from productivity, freeing the employee for greater intellectual output. If the computer is employed effectively, it is an asset. Improper planning for use of the computer will make the user a slave to the machine. Therefore, it is important to expect the machine to perform those tasks which it does best:

- Maintaining and searching records
- Storing greater amounts of data in less space than needed for paper files
- Doing tedious counts, statistical problems, and complex computations
- Returning selected, summarized information to the user quickly

Kinds of Data Captured. Tracking workplace hazards requires the collection of data on agents, conditions, and employees. Numerous other files can be added to the basic list shown in Figures. 1.9b and 1.9c.

Quantities of Data. A worthwhile lesson on data volume can be learned from observing the rate at which information increases. A record contains

COMPUTER USES

1. Office Automation

Word Processing
Spreadsheets
Computer-aided design
Graphics

2. Process Controls

Regulating process flows

3. Robotics

Lifting and moving objects
Automating processing

4. Data Logging

Recording repetitive data input

5. Decision-support

Expert systems

Fig. 1.9a Common uses of a computer.

many pieces of related data. A file is a collection of records. Begin with only 100 employees and five hazardous agents in a stable hypothetical plant. Each employee has one demographic record. Using conservative figures, we will estimate that these employees have an average of two work histories over a three-year period. The total number of records for the employees at the end of this time is 300, but if each record contains an average of twenty-five data items, the total is 7500 pieces of related information.

Each of the five agents has an MSDS with fourteen files. A record holds fifteen pieces of related data. Thus, one MSDS will cover 210 pieces of data, and five agents will put 1050 pieces of information into the database.

If each employee has an annual physical exam, and if seven tests (files) are conducted over the three-year period, there will be 52,500 pieces of medical data on hand.

There are ten accidents in the three-year period, each involving a First Report, an investigation, determination of cause(s), and corrective ac-

EMPLOYEE FILES

- Demographics
- Work histories

MATERIAL SAFETY DATA SHEET FILES

- Agent Description
- Hazardous Components
- Physical and Chemical Characteristics
- Reactivity
- Emergency First Aid
- Fire and Explosion Hazards
- Spill/Leak Information
- Health Hazards
- Toxicity References
- Protective Measures
- Environmental Effects
- Labeling Requirements
- Shipping Data
- Transportation Data

OCCUPATIONAL MEDICINE FILES

- Employee Clinic Visits
  Illness complaints
  Injuries
  Allergy and immunization shots
  Blood pressure monitoring
- Physical Exams
  Pre-employment (baseline) exams
  Periodic exams
  Audiograms
  Electrocardiograms
  Pulmonary function tests
  Xrays
  Vision tests
- Special Programs
  Wellness programs
  Blood pressure programs
  Back programs
  Fitness programs
  Hearing conservation programs
  Smoking cessation programs
- Medical Questionnaires
  Medical histories
  Health assessments
- Medical Equipment
  Inventory
  Calibrations
  Maintenance services
- Laboratory tests
  Blood chemistry
  Urinalysis
  Special tests
- Other
  Medical records location
  Medical referrals
  Work restrictions
  Mortality records
  Summary of findings
  Problem lists

Fig. 1.9b Some basic occupational health and safety records for the computer.

tion(s). This totals 1000 pieces of data. The safety manager conducted two surveys per year, for a total of another 150 pieces of data.

Twice each year, the industrial hygienist ran personal sampling surveys for two agents on each of the employees and six area surveys on each of the agents.

INDUSTRIAL HYGIENE FILES

- Surveys
  Air monitoring
  Noise monitoring
  Stack emissions monitoring
  Water monitoring
  Radiation monitoring
  Underground storage tank monitoring

- Laboratory Analyses
  Sampling analyses results
  Laboratory calculations
  Descriptive statistics

- Agent Inventory
  Agent storage locations
  Agent usage

- Exposure data
  Exposure assessments
  Sampling strategies
  Exposure calculations
  Exposure assignments
  Exposure histories

- Other
  Industrial hygiene equipment inventory
  Equipment calibrations
  Equipment maintenance
  Permits
  Waste management
  Shipments

SAFETY MANAGEMENT FILES

- Incident Investigation
  First Report
  Investigation details
  Incident causes
  Corrective actions
  Property losses

- Job Descriptions
  Job safety analysis
  Job fitness assessments

- Personal Protective Equipment
  PPE inventory
  PPE assignments

- Safety Programs
  Fire control programs
  Special programs

- Audits and Inspections
  Plant audits
  Safety inspections
  Safety alerts

- Training Courses
  Course descriptions
  Class rosters
  Training specifications

- Other
  Equipment inventory
  Ventilation data

Fig. 1.9c Some basic occupational health and safety records for the computer (cont.).

Each employee has received training for each job. That adds 5000 more pieces of data.

Without any additional records like clinic visits, job analyses, equipment inventories, personal protective equipment assignments, and assorted other files, it is interesting to add up the number of pieces of information on hand at the end of the three-year period:

Employee records	7,500
MSDSs	1,050
Physical exams	52,500
Accidents	1,000
Safety surveys	150
Area/personal surveys	30,750
Training	5,000
	97,950

Flexible Design. An occupational safety and health data collection is dynamic. Data requirements change as new standards requirements are added, new employees are added, new jobs become necessary, and new hazards are found. New data cannot nullify old data. Continuity of the database is important. A flexible database will accept changes by small degrees without corrupting existing data.

In a growing organization, the employee population may expand, new plants acquired, new products developed, and processes changed. A flexible tracking system will absorb these improvements. The computer processes the data, and users are able to retrieve and review the information in a timely manner by screen display, ad hoc reports, or standard reports. Automation is a primary tool for tracking workplace hazards.

1.10 EMPLOYEE AND COMMUNITY RIGHT-TO-KNOW

Right-to-Know. In the United States employees and the community have the legal right to know about hazardous materials locations and potential danger. For the company, this means that records must be kept updated, especially MSDSs, agent tracking records, and monitoring data. Unions scrutinize exposure data for employees, and environmentalists scrutinize emissions data, always poised to pounce should the company falter. Clearly, the government expects the company to control hazardous exposures, and the public expects to monitor the company's efforts.

Records for Right-to-Know. The first prerequisite for meeting this responsibility is a committed company management. Commitment entails persistent workplace monitoring and employee health surveillance, supported by thorough recordkeeping. Records are the company's best defense when charged with negligence. Contention invariably arises in the absence of good recordkeeping procedures or because of the total disregard for records. Believable records are the public's measure of a company's good faith, along with adequate employee and community protective measures.

In litigation, exposure measurements are the usual subject of contention. Expert witnesses are called upon for opinions about what is or is not in the records. Methods of measurement, their interpretation, and their reporting often constitute the major part of the testimony heard. Consistency in data and adequate frequency of testing provide strong evidence that the company was looking after its workplace hazards in good faith.

Company Aversions to Records. A common fear among companies is that records are potentially harmful because they will provide contestable data which, in litigation, may lead to a weakness in the company's case. However, for the most part good records demonstrate the company's best intentions because there is a preponderance of evidence that the company was working to meet accepted standards.

The news media rarely publicize the companies that make a conscientious effort to meet occupational health and safety standards. Those companies that flagrantly violate laws and concern for employee and community are highlighted instead. When citation hearings and court action take place, violators have nothing to blame but their own stubborn callousness for the penalties they receive.

Government, Company, and Community Working Relationship. Title III of the Superfund Amendments establishes methods for making the community aware of potentially dangerous exposures to chemical hazards. The company must provide a list of hazardous substances on site or the MSDSs to the local emergency planning committee. By law, this information can be shared with the community. Shared information should extend several benefits to both the public and the company, as shown in Figure 1.10.

The effect is to forge a working relationship between the government, company, and community for the control of hazardous substances and conditions. We need to reduce the adversary relationship between the community and the company in order to maximize economic preservation of the business community and personal protection for employees and public.

Company agent inventory and agent tracking records should furnish data required by Title III in the form of quantities of hazardous chemicals purchased, manufactured, processed, stored, transported, or disposed of. These data can be compiled nationally for better evidence of environmental control.

EFFECTS OF SUPERFUND AMENDMENT REQUIREMENTS

Requires preparation in the event of
an unexpected substance release

Gives the public confidence that the
company will control its hazardous
substances

Causes company to closely monitor
controls over substance release

Provides State and Federal governments
with national inventory of quantities
and locations of hazardous substances

Fig. 1.10 The effects of Superfund Amendment requirements.

Early Detection. Without extensive records on employee exposures and health, it is impossible to detect incipient harm to human health. A computerized recordkeeping system will provide the means for reliable, repetitive measurements and frequent review to detect early warning signs. Borderline and abnormal results will be reported to employees.

Well-kept, reliable records are essential to the company if informed employees and the community are to have confidence in company management. Typically, management chafes at government regulation, but in the case of employee and community right-to-know laws, a system of checks and balances creates more protection of our environment against the potentially devastating harm to everyone from exposure to chemical and physical hazards.

1.11 STRATEGIC PLANNING

Steps in Strategic Planning. In deciding how to automate an occupational safety and health system, fundamental company policy must be formulated and adopted by top management. Medical, industrial hygiene, and safety personnel can assist the company with strategic planning advice for a program that:

- Protects the employee in the workplace.
- Decreases the costs of employee health and safety.
- Contributes, wherever possible, to the profit margin.

The first step in strategic planning is for occupational safety and health professionals as a group to propose an integrated, computerized occupational safety and health program to the company. The initial points to be made are that:

- Regulations require the company to implement occupational health and safety programs.
- The company is the legally accountable party for government mandates from agencies like OSHA and EPA.
- A fully supported, integrated, computerized occupational health and safety program can protect the profit margin by increasing employee fitness, reducing compensation and disability expense, and avoiding regulatory intervention.

The second step is to set the stage for management to take control of occupational health and safety. Professional advice and expertise can be freely given by company specialists in occupational medicine, industrial hygiene, and safety, but management should not be permitted to transfer its final responsibility for employee and public health and safety. When occupational safety and health professionals accept the mantle of accountability, they find themselves in the weak position of forever going to management with a tin cup. In reality, their position should be that of a proactive watchdog, monitoring, surveilling, collecting data, evaluating results, recommending, and reporting.

The third step is to recommend and sell the plan for an occupational health and safety program to management. Unfortunately, most professionals start with this step, setting the stage for a tin cup instead of a watchdog future. If the first and second steps are carried out, the company will need advice and will turn to the specialists to solve their problems. Company occupational health and safety problems should be itemized, along with their possible solutions. Program approaches can be recommended, along with the resources needed and the alternatives. A vital resource is a reliable, fast recordkeeping system.

Company financial reports can be scrutinized for line items to be improved by an aggressive occupational health and safety program. If the company wants to reap the financial reward, management will have to make an investment in kind. Wherever possible, the cost of these improvements should be presented, along with ways that the program can save money. Arguments can be based on published data that exist in abundance, such as:

- Reduction of workman's compensation and disability costs
- Reduction of insurance premiums for the company
- Possible return on insurance premiums
- Increased productivity
- Reduced absenteeism
- Avoidance of government penalties

Use strategic planning to propose the program. Instead of presenting a program for the specialist to carry out, make the program one for the company to launch, one that emphasizes employee involvement and leadership. Encourage management to see the advantage of a partnership of the company, employees, and unions, with the common reward of safety and good health.

The Company's Responsibility. Basic to all these suggestions is the objective of having company management bear the responsibility for employee and community safety and health. The company that steps forward to assume this role will reap benefits.

Professionals in occupational health and safety are on-site consultants. Their role is to render professional service to the company, not run interference. When the professionals assume responsibility for the battle against hazardous exposure, they expose themselves to possible litigation. They can be held liable for professional negligence, meaning that they are at personal risk when performing their jobs. This is evident from the growing number of court cases against health and safety professionals. After all, it is safer for the plaintiff to take on one person than a whole corporation. Therefore, it is in the best interest of health and safety professionals to keep the responsibility in company hands and to maintain very good records.

1.12 SUMMARY

An integrated occupational health and safety system is a computerized database for shared information management that includes data on occupational medicine, industrial hygiene, and safety. This should be an integral part of the company's health and safety program to protect employees and the community from exposure to hazardous agents and conditions.

The integrated database offers cost-effective administration, reliability, and efficient retrieval and output. Data on hazardous agents and employee health can be shared on-line or by reports. System design can be based on a centralized or decentralized system under company or local user control. Joint decisions involving management, information services, and users will establish corporate goals for the system with allowance for future growth.

Cost-effectiveness analysis is preferred over cost/benefit analysis because of the difficulty of placing dollar values on human health and safety. Advantages of an occupational health and safety computer system to the company are protection of assets, a competitive edge because of improved production, defense against litigation, good public relations, control of disability and compensation costs, and defense against government citations. Both in the United States and abroad, a good health and safety program has proved to be good business. Automation effectively tracks hazards in the workplace; improves the working relationship between company, government, and community; and protects workers from harmful and costly exposure to chemical and physical agents.

It takes strategic planning to propose the purchase and development of an integrated occupational health and safety computer system for management's approval.

Chapter 2

MATCHING APPLICATIONS SOFTWARE WITH SUPPORTING HARDWARE AND UNDERLYING SOFTWARE

Before an occupational health and safety system can be developed, resources must be marshaled based on who uses the system, for what purpose, and with what available funding. In addition to the health and safety professionals who need to computerize their data management, there may be secondary users.

Along with operating and database management software, hardware will be needed to run the occupational health and safety applications software. Computer languages for coding software have grown through four generations and are evolving into a fifth generation. Fourth-generation tools save time and money in development by automating the programming process. Screen painters and report generators provide a means for fast development, and prototyping makes it possible not only for users to review the system during development but also to play an important role in shaping the final product for their needs.

Database design is a painstaking process that begins with the systems analyst and inspection of paper forms being used, and leads to a hierarchical, network, or relational database designed to avoid data redundancy. For ad hoc querying of the database, a query language is needed to give users the freedom to review data on-line.

The developer contributes the necessary systems expertise, while the users contribute the required applications expertise to form a development partnership.

2.1 WHO WILL USE THE SYSTEM AND HOW?

Primary Users. The primary users will be personnel in the medical clinic, industrial hygiene, and safety offices, who will be able to perform their jobs more efficiently. This will be like office automation. Information services will provide support for system modifications, maintenance, and repair.

All three principal user groups typically have high volumes of data input. As an example, the medical clinic collecting laboratory test data on 100 employees annually for blood tests, urinalyses, and hematology will find their paper files bulging within a short period. When these data are reviewed, the blood test will probably contain a minimum of 20 test results, the urinalysis may contain 10 results, and the hematology report may cover about 12 results. This is a total of 42 results on each of 100 employees for one annual physical, or 4200 test results. In five years, this would require analysis of 21,000 test results. Not included are test results from unscheduled or other scheduled clinic visits.

Secondary Users. Secondary users include risk management, medical and social researchers, and government agencies. While these users do not have firsthand access to the system, they need data from it. An example of a government agency use is the Bureau of Labor Statistics (BLS) and the information on the OSHA 200 log. The system would generate and print the OSHA 200 with all required counts and calculations for delivery to the BLS.

Medical and social researchers might be epidemiologists within a corporation or elsewhere. For research and epidemiology, subsets of the data must be drawn from the base system for statistical analysis with software like Statistical Analysis System (SAS) and Statistical Package for the Social Sciences (SPSS). In research it is difficult to obtain data that are reliable for statistical purposes, and in a corporate setting it is important to have compatible data from all comparable plants. A company epidemiologist should be consulted when setting up and changing the database to ensure good results for future studies.

Section 1.8 discussed the concerns of the ILO for occupational safety and health and the active concern of many countries with obtaining information on hazardous substances in the workplace. There is a vital need for data from all American plants, regardless of size or kind of industry, to be contributed to central data banks. In this sense, there is international interest in an integrated occupational safety and health system. This is another example of secondary use of the computer system.

Reports. There is a critical need to mobilize data for retrieval at any time. With a staff working on tight schedules, it is burdensome to extract data from paper files and organize them for analysis. A computer system provides rapid data retrieval and is cost-effective. Depending upon the size and configuration of the system, a batch run done overnight makes efficient use of both staff and computer facilities to display or print data grouped by age, job code, sex, or any other category. Batch processing is usually done on off-hours and runs a group of records all at once. Summary data can be obtained, such as averages, percentages, and ranges. When placed in a data subset, this information can be transported to SAS, SPSS, BMD Biomedical Computer Program (BMDP), or some other statistical package for rigorous statistical analysis.

Data are not computerized for the sake of input. Computer storage and processing time cost a lot more money than paper files. The advantage of the computer is that the data can be processed to produce timely information in a cost-effective manner. Processing can be as simple as storing the data or as complex as performing calculations on the data before or after storage. The data may need some kind of conversion before storage or usage, or they may require comparison with other data. These are structured tasks that are repetitive and time-consuming for humans but speedy and effective for a computer.

Computer data can be retrieved in several ways, either separately or in combination:

- Queries
- On-screen displays
- Printed reports

Queries are generally ad hoc requests for specific data. The computer is programmed to accept user commands for this type of retrieval, the standard query language being SQL (Structured Query Language). (See Section 2.6. for a discussion of SQL.) If a query is reusable, it is also possible to "can" it for repeated use, which eliminates the need to enter it each time. Canned queries are saved and stored, and each time the user requests information, the most recent data are included. Once mastered, SQL is a powerful ad hoc information tool. Ad hoc data retrieval can be described as "taste-testing," after which data can be subset for analysis.

Programmed reports can be previewed on the screen for the user's approval before sending them to the printer. When programmed, the reports can be simple or complex in design, and of any desired length. Simple reports can be straightforward presentations (dumps) of data in the database. They show lists of information, which may be sorted according to the needs of the user. There may be headings, columns, and totals. Subtotals and breakdowns of the data (by department or job code, for instance) are more typical of complex reports.

Complex reports show breakdowns within groups. Subtotals are followed by final totals at the conclusion of the report. There may even be groups within groups, like job codes within departments within different plants. Complex reports may have multiple parts, such as a section on incident investigation data and another section on incident cost breakdowns.

Simple and complex reports should be selectable from screen menus, giving the user the opportunity to request the report needed. There are software packages available for "writing" reports. Many users will not take the time to learn how to use the report writer, preferring to have information services write reports according to their specifications.

Professional users, especially users with large volumes of technical data, employ assistants to record data and search records for information. Given a computer, these professional users will not necessarily employ fewer assistants, but the assistants will perform their tasks more efficiently and be able to supply the professionals promptly with comprehensive information.

In the daily routine, the real users will be the assistants who will enter data, correct them, and retrieve them as necessary. They will be the primary beneficiaries of a *user-friendly* system, one that eases communication between human and machine. A basic requirement is an easy-to-follow menu system with readily understandable prompts and messages. User friendliness is of the utmost importance for a user who lacks computer experience. These users will be impatient with a system that is difficult to understand or that forces them to depend on information services for frequent help. A system that does not perform smoothly and efficiently for daily users soon is scrapped in favor of a less exasperating and more familiar paper file.

2.2 WHAT KIND OF COMPUTER EQUIPMENT IS AVAILABLE?

Computer Cost. The most expensive system is the mainframe and the least expensive is the microcomputer, but the question to be addressed is, how much computer power can you buy for your dollar? The greatest power per dollar is in the microcomputer due to the significant progress in recent years in microchip development, which provides increased computer memory and speed. Microchips are integrated electronic circuits that are microscopically etched on silicon wafers and linked together to make a computer system. At the same time that microcomputer chip development has vastly increased computer power, computer system cost has gone down. This is true for all three sizes of machines. Minicomputer power is very adaptable to varying needs because modular construction permits users to increase the size of the system as needed.

In addition to the cost of the computer hardware, consider the necessary space. Microcomputers and some minicomputers sit on desktops, while mainframes require room space and possibly air- conditioned accommodations. Minicomputer modules can be stacked up like blocks beside an office desk.

Mainframe costs are greatest because these machines need trained operating personnel. Microcomputers can be operated by one individual with limited training. Minicomputers are adaptable to multiuser situations, which may make them cost-effective on a per person basis, but they require training to operate.

Computer Capacity. Memory and storage are two measures of computer capacity. *RAM* stands for *random access memory* and means that data can be accessed randomly. When a program is activated, it is placed in RAM so that as data are entered, they can be processed as directed. This is why large capacity memory is necessary in order to accommodate an integrated program of the size needed for occupational health and safety.

Memory capacity has been immensely increased in recent years for all three sizes of computers and provides great computing power. For its size, the microcomputer has more memory capacity than the minicomputer or mainframe.

Storage capacity is important to the health and safety computer system. Storage is the part of the system where data are preserved. Capacity needed is dependent on how many records need to be kept currently available and how fast they accumulate. When storage is used up over time, the older records can be archived in auxiliary storage. Archived records are those that are not currently needed and can be conveniently separated from the computer for long-term preservation. Auxiliary storage for microcomputers is on diskettes and tape cartridges; and on the larger machines it is on disk packs.

Microcomputers for occupational health and safety will need hard disk drives to accommodate the operating storage load. Modularization of minicomputers provides an opportunity to increase storage as required, while mainframes have a large initial storage capacity. In environments with large volumes of current data, a large amount of storage is necessary. A plant with occupational health and safety records for 5000 employees needs substantial computer storage, like that available in a mainframe, while records for 50 employees would work well on a microcomputer with sufficient storage capacity.

Although storage is no longer a limiting factor in a computer system, a caveat should be added: Indiscriminate use of storage is to be avoided, especially on microcomputers. This is particularly pertinent to unrestricted use of text files that clutter up space. Text is a highly desirable feature in the occupational health and safety system, but it should be appended to explain factual data, not be used as the data.

Virtual memory systems are available in larger computers and are advantageous for applications software. In a virtual memory system, the operating software directs the processor to shift segments of the program not in use to a reserved direct-access area. As information is needed for processing, it is recalled; at the same time, temporarily unneeded information is sent to the reserve area. This gives the user the impression that memory capacity is unlimited. The advantage is that applications programs can be very large, but the disadvantage is that retrieval takes time, slowing down program execution and increasing system response time.

Operating Systems. Operating systems differ with the machine manufacturer. IBM and compatibles use the disk operating system (DOS) or Unix; Apple and Macintosh have their own operating systems. A machine-dependent operating system coordinates all the operations going on within the computer. Despite microcomputer memory improvement, DOS operating software has not kept pace with new microcomputer applications. That problem has been overcome by new hardware technology and operating software called OS/2. OS/2 is the operating system for the new IBM computers that handles operations beyond the limitations of DOS. DOS limits the microcomputer to single-user environments and single program operation at any one time.

Video Display. A video display terminal (VDT) is the viewing screen that is linked to the computer machine. This is often called a *monitor* because it is the user's "window" on what is happening within the computer. A component is the cathode ray tube (CRT), which is like a picture tube in a television set, except that the computer tube has been modified for the display of alphabetic and numeric characters rather than pictures. When a keyboard is added to the monitor, the combination is called a *terminal.* Terminals are wired to minicomputers and mainframes over distances of many feet throughout a building or group of buildings. A keyboard and monitor cabled to a microprocessor only inches away becomes part of a microcomputer system.

Video display screens give viewers an opportunity to avoid eyestrain by selecting screen colors that suit their taste. Monochrome monitors are available in white on black, green on black, and amber on black combinations. Intensity controls permit adjustment for eye comfort and brightness of characters. When preferred, the combinations can be reversed. Monochrome monitors are commonly used in office environments.

Color monitors are available for smaller computers and are a matter of personal preference. They are especially helpful when using graphics, but they are more expensive than monochrome monitors. Some people find them difficult to watch on a daily basis because of eyestrain.

The use of graphics with the microcomputer system will require a high-resolution video monitor. Graphics capabilities may require the purchase of graphics cards on some machines in order to accommodate the high screen resolution desirable for graphics display. Color/graphics adapters control color monitors and provide high-resolution graphics. They are relatively expensive.

Configuration. Computer machines are configured for user preferences, meaning that the user has some control over the organization of the computer system and options that can be made for its use. For instance, video display can be done in the usual mode of bright characters on a black background, or in reverse with black characters on a bright background. Memory and storage devices can be chosen to suit user needs, and peripheral equipment like printers can be selected to match the user's requirements. Peripheral equipment includes printers, plotters, and modems. Dot matrix and laser printers are typical choices for microcomputers, although ink jets and daisy wheels are also available. High speed printers for the larger machines vary in cost with increasing speed and output volume. In addition to speed and clarity of print, printers are selected for different widths of paper and the ability to feed in single sheets of paper. An important aspect of selecting printers for occupational health and safety is determining the desired size of paper for reports.

Combining Systems. Information systems will offer advice on the possibility of creating multiuser systems that combine machines for maximum economy and user efficiency. They may also recommend a computer network. A network can make workstations of microcomputers and link them to each other and/or to larger computers. No longer are microcomputers islands of information. Workstations are integrating data management. The microcomputer becomes a front-end processor for the system, while the minicomputer or mainframe becomes the main storage facility. This permits the larger machines to serve multiple purposes and makes for cost-effective computer power. Because of the many demands that an organization makes on mainframes and minicomputers, performance and response time are slowed, which is why the role of microcomputer workstations is expanding.

Sharing computer resources is the trend at the beginning of the 1990s. This includes shared hardware, software, and databases, and has been made possible by microchip improvements for increased memory. The expanding market for the personal computer has also played a large role in the move toward multiuser systems. Systems that accommodate up to

15 or 16 users are highly adaptable to an occupational health and safety integrated computer system. Where users have their own desktop microcomputer, it is possible to link these machines into a single system, which is a cost-effective enhancement to information management.

Newer software has been developed to support the multiuser integrated environment. IBM's OS/2, OS/2 Extended, and Xenix are the principal contenders in this line, since the old DOS operating system was unable to support more than one user at a time. These new operating systems also support multitasking, which permits two or more software programs to be run simultaneously. Multitasking is another capability that requires ample memory capacity.

Insertion of a link circuit card in the microcomputer provides linkage to minicomputers and mainframes, immensely increasing computer power. Current data can be used in the workstation environment and transmitted to the minicomputer or mainframe for archiving when required.

Networks are the more expensive way to create a multiuser environment because of the hardware and software needed to link the users together. Cabling is expensive, and a network may be as much as 50-60% more costly than a simple multiuser system, but it will handle up to approximately 60 users connected to one host computer. The host computer is the central processing machine to which everything else is linked. The differences between multiuser and networked systems are few, and the user will see them as both operating alike. The differences are chiefly in the operating software and hardware connections. A network may incorporate several uses, while the multiuser system is designed for a specific use.

Advantages and Disadvantages. The advantages of microcomputers for occupational health and safety are that they are the least expensive computer system and are more suited for department use because the user controls the system. On the other hand, minicomputers are faster, more efficient, and can be modularized, handling larger volumes of memory and storage. The mainframe is built for very large capacity and can handle accumulating data from several plants, which is ideal for epidemiological analysis.

The disadvantages of microcomputers are that they do not support high-volume printer output and that unless they are on a multiuser or network system, it is a nuisance to transport data on diskettes or tapes for integrated use. Minicomputers and mainframes are more expensive than standalone microcomputers, and they require operating and systems management personnel. Systems management personnel are also required for multiuser and network systems.

Users of minicomputers and mainframes can be disconcerted by their slow response time. Response time is the period that the computer takes to transmit directions from the user's keyboard to the computer, process them within the computer, work within the database records, and transmit the answer back to the video terminal. When many users work on a minicomputer or mainframe system, they can slow the response time significantly and frustrate users.

For smaller companies, the mainframe is not economically feasible. In such cases, the minicomputer is a likely compromise between the limitations of the microcomputer and the expense of the mainframe.

2.3 WHAT IS APPLICATIONS SOFTWARE?

Applications software is a complete, usable set of programmed instructions designed to manage information for a specific user purpose. It is typically written in a high-level programming language that befits the kind of data required by the user, such as COBOL, FORTRAN, C, or Pascal.

Generally, applications software will run only on the same make of computer on which it was written. This is because the software is compiled, meaning that the programming statements are converted to machine language. If the applications developer wishes to make the software portable over more than one type of computer, it will have to be specially adapted to make the change. Software is described as *portable* when it can be run on more than one kind of hardware. Porting from IBM microcomputers to Apples would be prohibited because they are entirely different machine systems. Porting between more closely related microcomputer systems could be done by making relatively minor changes. When the compilers are compatible, porting between minicomputers and microcomputers is also feasible, even though it may be necessary to scale down the memory demands for the microcomputer.

In developing an application, a specific application objective is first outlined. For instance, an objective in an integrated occupational health and safety package might be to manage all the data that apply to employee exposure to hazardous substances. This might not include the requirements for epidemiological analysis of the data. In the plant or corporate environment, if information services planned to design and develop an occupational health and safety system, the first objective would be to work with the users on defining the purpose of the applications software and its limits. If the objective of the package is to manage data on employee exposure, the software should not be expected to carry out ventilation modeling calculations.

Every user would like a desktop computer and the capability to handle all the information tasks required during the day. This is a natural wish, but the applications would need to be separate, since they may utilize very different types of data, and processing may range from simple filing to complex statistical calculations. It is not yet reasonable to integrate all the tasks confronting a user into a single software package. This book is about integrating a system across users of different disciplines in order to provide information on one defined subject. The limit on the information collected can be confined to the employee and the environment, or it could extend to the employee, the plant environment, and information for community right-to-know issues, which could include waste streams beyond the plant site.

Typical applications software is menu driven, so the user can select options from the screen rather than learn a command language. Menus also provide a flexible method for navigating directly through the system to areas of interest. This makes all parts of the system easily and quickly accessible.

Developers of applications software depend upon two types of experts within their organization: computer specialists and specialists in the field of the application. The first contributes specifications for the design and applications requirements that enable the user to use a computer; the second contributes specifications for making the computer carry out the intended purpose.

2.4 WHAT ARE 4GL AND 5GL?

Software Development. Software development is labor intensive. It must be carefully designed, modeled, programmed, and adequately tested before it is a usable product. The ideal method is to automate as much of this work as possible. Progress has been made toward this end, but more needs to be achieved.

Machine Language. Software is the computer language product by which the human directs the computer. When computers were developed in the 1940s, each movement of electrons in the circuits had to be anticipated by the programmer and set by bits, the binary system of switching tiny circuits on and off. This was programming in *machine language*. John von Neumann is said to have vividly described this tortuous, error-prone manner of communicating with the computer machine as an extraordinary human effort in which the result depends on a sequence of a billion steps and each step really matters. It was von Neumann who proposed that the computer machine be made to store a program in numerical form. Prior to that time, thousands of switches and wires had to be manipulated to carry out the program. A computer language that uses a numerical method of programming is termed a *first generation language*.

Assembly Language. By the 1950s, it was evident that the programming process demanded an automated means for human direction of the machine, and the *second generation* of programming language evolved. The machine was set to receive coded mnemonic statements in *assembly language* from the programmer, and the machine then converted these statements to machine language. By the end of the decade, machine chips each had their own language. This was a vast improvement over hand setting each bit, and it reduced the error potential as well as saving programmer time. However, each developer produced a machine with its own internal design and, therefore, its own language. The uniqueness of the machines required unique languages, none compatible with the others.

High-Level Languages. Following assembly language, *high-level* languages were developed, which used English statements rather than codes. Precisely worded commands are converted back to assembly language before final conversion to machine language. The converter is a special program called a *compiler,* which is a high-level program in its own right that converts a high-level language to machine language. Compilers accept only precisely correct commands for conversion, thus requiring skilled programmers.

In the 1960s software was of two types: programmed instructions for operating the machine and programmed instructions for carrying out the application. Operations software controlled the machine's resources, such as directing electronic traffic, handling memory, accepting input data, carrying out calculations, sending data to storage, retrieving data for output, and sending data to the VDT or printer. The burden of directing machine actions was removed from the user. Applications software, on the other hand, allowed the user to dictate to the machine how the data should be used. It became the interface between the human and the machine operation. The chief software problem has been software reliability because of the high potential for programming errors, but from the 1960s on, the demand for software production became so great that software reliability became a primary issue.

Programming. Programming is a painstakingly slow development process that is never apparent to the public. John Q. Public only knows that the computer did not credit his last payment; John is infuriated at the programmer who couldn't get his act together and the programmer agonizes through thousands of programmed statements to find the few characters that produced an error in John's statement. Computer reliability depends upon programming perfection. To achieve this, we need simplified programming and a reduced possibility of human error.

By the time that microcomputers emerged in the 1970s, both hardware and software development were focusing on user appeal. By the 1980s, a desktop machine gave the user a sense of being his or her own master, rather than being at the mercy of a faceless programmer. "Do-it-yourself" was a concept that bubbled over into the computer world. In the mid-1960s, the idea of having tools with which to do programming took shape. These tools automated many of the repetitive tasks in programming and thereby reduced the number of errors. They also allowed users to do their own development. The next decade focused more on hardware improvements than on software, but by the 1980s, software development had become such a critical factor in the growth of the computer industry that the time had come for significantly improved programming.

Definition of the generations in software development is not as easy as in hardware development. In recent years, newly developed databases and hardware have been described as fourth generation, and somehow software assumed the same phase designation, but a good definition of fourth-generation software is elusive. One of the early programming languages was COBOL, a highly procedural mode in which the statements are ordered in a specific operational sequence. Writing applications software consumes a lot of programming time and computer memory. Some people envisioned the possibility of a language that would be specific to an application and issue a single command governing an entire procedure or set of procedures. In essence, these languages would be programming tools. The benefit would be time savings and increased reliability. They acquired the name *fourth generation languages (4GL)* and were nonprocedural, their most intensive development having taken place in the 1980s. The earlier languages were all procedural in that they were command driven, instructing the machine where to go to find a file and which path to take to get there. Nonprocedural languages do not specify in detail how to carry out a request, leaving the system to determine the path to be taken.

The 4GL tools include screen painters, data dictionaries, macro-processors, report writers, and some query programs. In some cases spreadsheets and decision support systems are also included. They are user friendly, and many are aimed at the nonprogramming end user who is bent on producing software uniquely adapted to a specific task.

We are now developing *fifth-generation languages,* which are part of the artificial intelligence concept. Expert systems make use of this technology which attempts to emulate human decision-making processes. The medical world has already experimented with diagnoses. Safety managers have been offered small, special purpose microcomputer systems. One example is a system for deciding which brand and type of personal protective glovewear should be selected, given the agent and chemical conditions for employee exposure.

Expert Systems. Expert systems assume that reasoning is a series of logical steps that can be sequenced in the computer. The user sets up a problem to be resolved or a question to be answered, and the system responds with a series of logically organized questions, using a fixed set of rules. Each reply by the user prompts the system to select the next question path until, all selection paths having been exhausted, a conclusion emerges. The key to expert systems development is the human expert who breaks the problem down into logical steps for programming solutions. In solving problems, however, humans tend to bypass some logical reasoning, given the insightful ability to skip from A to E without going through B, C, and D. The computer lacks this ability and must be programmed to step through B, C, and D before tackling E. Thus, fundamental to expert system development is the ability of the human expert to define his or her own thought processes. Systems developers expend immense amounts of time "picking the brains" of human experts, attempting to exclude intuition and opinion from the processing steps because the solution demands cognitive skills, not common sense.

An expert system has a database of facts about the subject. Selections are made from the database, depending on the data supplied by the user. Once a number of early possible paths have been sorted through, there is a narrowing of possibilities pointing to a solution. Sorting through possibilities is a natural analytic capability of the computer. The computer can also store model sets. Once displayed, the user can select the model that most closely fits the given conditions. This also sets the computer path on a solution course.

Expert systems are difficult to produce because they require intensive examination and input of humans in order to follow paths of human reasoning. Very few are in use as we enter the decade of the 1990s, but a lot of research is being done in this field because of its potential for expanding computer applications in the coming century.

4GL Tools and Occupational Health and Safety. For occupational health and safety, 4GL tools hold special promise. One physician may use the Snellen system for measuring visual acuity. Another may use a numerical system that equates to Snellen measurements. The vision tests are alike, with the exception of the field lengths on the screen. The basic computer program remains the same. A screen painter can generate the screen for either kind of data entry without requiring a programmer to touch the underlying program. This saves time and prevents the possibility of making programming changes that would threaten the reliability of the program.

A report writer permits the physician, industrial hygienist, and safety manager to structure the computer printout according to the needs of the company. The report of one month may be repeated next month, and when it becomes evident that the information needs revision, the report can be modified or a new report can be designed and produced by "painting" a new format or changing the data to be retrieved.

The screen painter is similar to a word processor in that characters can be moved on the screen at will. Each data input item becomes a variable which the user can define for the data dictionary, a catalog of all the variables and files in the system with their attributes (characteristics). In addition to the name of the variable, its length, and the kinds of characters it will accept, there may be special conditions requested by the user, such as checking the range of a number entry or using the entry in a calculation. The data dictionary may operate with a DBMS, which may be either a separate system that overlays the applications program or part of the program itself.

Advantages and Disadvantages of 4GLS. Until 4GLs were developed, integrated occupational health and safety software was at a great disadvantage. The data requirements keep changing as the discipline matures. For software written in a procedural language, changes necessitated tedious reprogramming in the affected areas. Not only was this costly in terms of programmer time, but reworking operable code increased the possibility of code error. After a few changes, patched code became extremely difficult to maintain.

With a screen painter, changes to the database or rearrangement of the screens may be carried out by the 4GL system itself, thus avoiding programmer interference. The more flexible the screen painter, the better the provisions for modification. With a screen painter, a data dictionary, and a report painter, the changed system is quickly available and ready to be tested and implemented in a fraction of the time required by a programmer. Unfortunately, there is a price for automated programming. The disadvantages of 4GLs are as follows:

1. Mechanically devised code is highly reiterative. For each command, there may be more lines than if a programmer were to devise an elegant method of accomplishing the same procedure with fewer statements.
2. Special programming features that may be appropriate to an application may not be achievable with a standard screen painter or

report writer. To add these to the code generated by the 4GL may require programmer intervention. This can be time-consuming.

3. Despite their user friendliness, 4GLs can slow computer response time and increase computer overhead.

It is no longer necessary to have obsolete occupational health and safety software. The older systems, which typically took years to develop, were out-of-date before they were completed because they were custom built. Automated generation of code saves time, money, and programming resources. It also allows the user to view the changes as they are made in order to ensure correctness.

Trends in Application Generators. One trend in application generators is to have versions for microcomputers as well as minicomputers and mainframes. Development can take place on the microcomputer, reducing the development demands on the larger machine. Other trends are to maximize the usefulness of the generator by putting together a screen painter, menu painter, data dictionary, database manager, and import/export utility. An import/export utility is a special program that permits data to be brought into existing files from another source or exported as files to another source. The user only needs to supply the data. In a larger system the application generator may set up a system that works over a separate database manager.

In designing an occupational health and safety system, both users and information services will be impressed by powerful screen painters and report painters because of their cost-effectiveness and flexibility. Information services will render the best advice about an appropriate DBMS.

Expert Systems for Occupational Health and Safety. Expert systems will be used by occupational health and safety personnel in decision-making tasks where there are a database, data analysis capability, possible models, an interactive display for communication between the user and the computer, and the expert system. The present trend in microcomputer expert systems in this field appears to be related to disaster modeling. If there were a hazardous chemical release with given wind conditions, for instance, what would be the effect on the surrounding locale at varying distances from the emergency site? What would be the possibilities for fire or explosion? When and where would evacuations be necessary? The expert system would provide graphic illustrations of affected areas and recommend emergency responses intended to contain the release, or to limit losses with timed analyses on the spread of the disaster. Depending on wind speed and direction, new models could be projected with damage assessments for each. Comparison of all the information generated by the system would lead to recommended strategies and company preparedness with adequately allocated resources.

The corporate advantage of such a system is that the user can query the system by changing the parameters, and the program can be rerun for as many agents as may threaten the community or any combination thereof. Such an analysis can provide broad information of value to all levels of management for discussions throughout the company.

The National Safety Council sponsored the development of a microcomputer expert system to interpret the Resource Conservation Recovery Act (RCRA) (Laumer and Shipman 1987) rules in order to determine the level of waste generation in a company and which permits would apply. In the occupational health and safety field, there are numerous areas that lend themselves to expert system analysis.

2.5 WHAT DATABASE CONDITIONS SUPPORT AN INTEGRATED SYSTEM?

A Database. A database is a collection of facts and figures that provides important information on a subject. A well-organized database will provide more information with a given amount of data than a string of conventional files. For any organization, the efficacy of the database is the degree to which it keeps the company informed about its business.

An Integrated Database. An integrated database may comprise several or more areas of information that give a comprehensive picture of the company's condition. The facts and figures are organized into related files. Typically, a file will contain data items taken from the printed form normally used for recording data.

A well-designed integrated database will screen out redundant (duplicate) data. For instance, in the occupational health and safety database, there will be information about the employee, the employee's work history, the employee's medical records, hazardous agents the employee may have been exposed to, accidents the employee has experienced, and training given the employee. The common elements in all of these data are the employee identification number, name, and social security number.

To repeat the employee information on every record would consume a lot of storage space and risk error. Every time the name is spelled or the numerical data are entered, there is the chance of entry error; and if there is a deletion or a modification of the entry, this information has to be changed in every location in the database in order to maintain consistency. This would be not only time-consuming, but error prone. Employee profiles or reports drawn from the database must be consistent and reliable. The database is an invaluable resource to be protected from human and machine error in every way possible.

Database Structure. Information services will design and construct the database, working from information supplied by the users. If the occupational health and safety system is be centered on the employee, the employee name, ID, and social security number will be the primary (key) fields in the database. Other data will be grouped in files linked to this principal information. For instance, there may be files on work histories, hazardous agent exposures, and on-the-job accidents, but each will carry the basic employee data as well. A basic method for linking the related files to the original employee information will be devised.

Typical types of database structure used in computer systems are:

1. *Hierarchical.* In these systems, a single piece of information, such as the employee's name, can provide access to the next level of information, such as job category, and this level may lead to a file of job performance ratings. The hierarchical system is always ordered from the most important piece of information about the subject to the least important piece of information.
2. *Network.* This structure is more complex. Data items in the files are linked so that navigation through the database is structured. No one item has a greater hierarchical position than another. Hence, navigation may begin anywhere on the network and follow a continuous path. The disadvantage is that it takes a lot of computer overhead to track the paths, and it requires considerable computer memory.
3. *Relational.* This structure is now preferred over the first two. The basic structure looks like a series of tables (matrices) of data. A table is the equivalent of a file; a row is the equivalent of a record; and a column is a field (data item) in the record. Records can be combined through a common element (e.g., the employee's social security number) to link together rows of the tables and furnish large amounts of data on an employee.

A Basic Design for Occupational Health and Safety. Most occupational data lend themselves well to database design. Data on the employee and on agents are the two key types of data required. Occupational medical personnel will be directly concerned with the employee's medical data, but there will be occasions to seek information about hazardous agents.

The industrial hygienist will work primarily with agent information, with occasions to apply agent exposures to an employee. Safety management will study employee assignments and job fitness, track employee training, and record incident investigation data. The central focus will be the information on employees and agents.

The database may be divided so that each type of user accesses the area of his or her specialty. A partitioned database is more efficient because it limits the combinations of data to be accessed, focusing on a specific kind of information to be obtained. Partitions should reflect the natural divisions in occupational health and safety such as employee data, medical data, agent data, and incident data. Specific to employee data would be demographic information, work histories, employee exposure records, and training records. The medical data are best partitioned off to ensure record confidentiality, despite the fact that they are also specific to the employee.

Database Processing. Processing a database is a special and expensive task. Special software used for the purpose will require higher operating costs in demanding more memory and CPU. The cost of mainframe DBMSs could be as much as an additional $250,000, but most present systems now sell for far less, and many are even available on microcomputers as well.

DMBSs have the ability to create and store large numbers of records and to edit, modify, and retrieve data easily and efficiently. Their greatest contribution is to format the records and the output for ready storage and access. They can process data by sorting, process calculations before storage or upon retrieval, and check data entry for accuracy. Security can be maintained through passwords to screen users for browsing and update access. They provide data dictionaries and on-line user help aids. Generally, they can be made to interface with other databases through special programs in order to place data in files or send data to files in other programs. This makes it possible to bring in a company employee database and keep it updated. It also makes it possible to export data to statistical packages for analysis.

The DMBS and Information Services. Most companies already have a DBMS that fulfills many of their database requirements. The DBMS must operate with the hardware, its operating system, utility programs, and applications software. The user interfaces only with the applications software, so the DBMS is transparent to the user. In other words, while the user concentrates on directing processing through the applications package, all of the database processing is independent of user direction. The only time the user is in direct communication with the DBMS is through use of its built-in tools like SQL. (See Section 2.6.)

When an occupational health and safety package is purchased, it must either be compatible with a known DBMS package or have its own built-in DBMS. If a company uses its present DBMS and develops an in-house occupational health and safety application, development must conform to the limitations or capabilities of the existing DBMS. It is not always safe to assume that any DBMS will handle occupational health and safety data. These are principally scientific data. Some DBMSs are better equipped to handle financial data.

Most information services departments are reluctant to buy additional DBMSs that will serve only one type of user because of the initial cost and additional expensive demands upon the system. Additional burdens on existing computer facilities may force the purchase of additional memory or even the addition of another computer to increase the CPU. Hence, a DBMS to accommodate occupational health and safety software is a significant consideration.

Not all DBMSs will handle large volumes of records. If the company has 1000 employees, the number of records in an integrated occupational health and safety system may accumulate very rapidly. Information services can project possible numbers of records in some period of time, perhaps a year. These specialists will quickly determine what demands will be made upon the DBMS.

2.6 WHAT IS SQL?

Database Output. The primary purpose for storing volumes of data is to retrieve them and supply information upon request from the user. The information is placed in some output report form, either on a VDT or on a printed page. There can be standard reporting formats for the information, or queries made to the database to yield an ad hoc response.

Ad Hoc Queries. Ad hoc querying of the database is one of the most useful features in an occupational health and safety computer system. It is done through a nonprocedural language called *Structured Query Language (SQL)* developed for database use. This evolved from IBM research and development during the 1970s. SQL operates with relational databases and is usable only as a database language. Vendors have elaborated on the initial query system. A proliferation of query systems created some chaos that was eventually addressed by the American National Standards Institute (ANSI). ANSI responded by setting some basic standards for SQL, leaving it to the vendors to contribute their own variations. Many vendors have chosen to adhere closely to the IBM SQL. This has made it practical to interface SQL, database managers, and applications software.

SQL can be used with programs written in procedural languages such as COBOL, FORTRAN, Pascal, and C. This is referred to as *embedded SQL*. Embedded SQL makes it possible to use the query language within applications programs written in procedural languages.

SQL Features. The DBMS supports SQL with a data definition language (DDL), a data manipulation language (DML), and a data control language (DCL). DDL creates and defines the relational tables. DML manipulates the table rows by retrieving them in their entirety or only certain columns, inserting information within a row, updating rows, or deleting rows. The DCL makes it possible for the database administrator to grant or revoke database privileges for users. A user may be permitted only to retrieve information or, additionally, to update and insert new rows.

Another DMBS support for SQL is the optimizer, which determines the optimal path to take in retrieving data from the many tables. Indexes are files sorted and ordered on a field other than the primary key. In one instance, the optimizer may determine that the requested information may best be found by indexes; in another case, a different search strategy may be more efficient. The optimizing feature is totally transparent to the user but is essential to the efficiency of SQL.

SQL Constructs. There are only about a dozen SQL constructs (commands), and these are simple English words. For professional users who prefer the freedom to search a database without depending on programmer services, the brief time and effort it takes to learn SQL is worthwhile. A well-written manual that comes with the application software should provide SQL instruction. The other convenient user reference is a comprehensive printout of all the files (tables) and the fields (variables) within each file. From this reference, the user will obtain the database names for each table and variable.

Ad hoc querying of a database is valuable to the professional user because:

- It permits direct on-line access to the database for immediate information. Simple questions can be readily answered without having to depend upon programmers or computer personnel.
- SQL supports simple descriptive statistics to be used with the query. In this manner the constructs Minimum, Maximum, and Average can be used to scan numerical data.

- Later statistical evaluation will benefit by preliminary scrutiny of the data. Ad hoc queries offer the user the ability to view data in different combinations before determining the optimal combination. After the user selects the best subset, SQL can create and save the file for subsequent use in statistical analysis software.

Illustrating SQL. Learning to use SQL is not as demanding as learning to program in a high-level language. With only a little practice, the user quickly becomes familiar with the structure of the query language and how it facilitates data retrieval. To encourage users to learn query by SQL, several pertinent illustrations will be given. Although medical queries will be used as examples, it would be just as easy to examine data for industrial hygiene and safety. We will begin with simple database queries and add specifications to show how SQL freely navigates around a relational database.

The primary commands are Select and From. *Select* refers to the variable(s) to be retrieved, and *From* refers to the table(s) in which they are located. Tables consist of columns and rows. Each variable in a file is represented by a column, which is headed by the variable name as given in the data dictionary. Each row of the table is a record. An employee table named EMPFILE may have many records. The example in Figure 2.6a shows that EMPFILE contains the variables Employee ID, the Employee Social Security Number, and the Employee Name. A real Employee Demographic table may also include the birth date, home address and telephone, spouse information, person to be notified in an emergency, hire date, termination date, and other related data. Each employee has a record, which is one row.

The user's reference manual says that Employee ID is labeled EMPID, Employee Social Security Number is labeled SSN, and Employee Name is labeled EMPNAME. We now have enough information to construct a simple SQL query, as in 2-1 of Figure 2.6b. The system would respond with the screen display in 2-2.

Obviously, the user is more interested in meaningful medical information. The illustration given simply shows the fundamental manner in which SQL commands are given for data retrieval. We will navigate

Employee ID	Employee SSN	Employee Name
11111	111-11-1111	Smith, Joseph L.
22222	222-22-2222	Martin, Sally A.
33333	333-33-3333	Feldman, Jonathan T.
44444	444-44-4444	Byrd, Matthew B.

Fig. 2.6a An example of retrieving an employee listing through SQL.

```
SELECT SSN, EMPNAME
                                                    2-1
FROM EMPFILE
```

```
    SSN               EMPNAME

111-11-1111     Smith, Joseph L.
222-22-2222     Martin, Sally A.                    2-2
333-33-3333     Feldman, Jonathan T.
444-44-4444     Byrd, Matthew B.
```

```
SELECT EMPID,LOCATION,CREATNIN

FROM EMPFILE,JOBHIST,LABDATA
                                                    2-3
WHERE EMPFILE.EMPID=LABDATA.EMPID

      EMPFILE.EMPID=JOBHIST.EMPID
```

```
EMPID         LOCATION        CREATNIN

11111           B101            1.2
22222           B101            0.9                 2-4
33333           B101            0.6
44444           A99             1.0
```

```
SELECT EMPID,JOBSD,CREATNIN

FROM EMPFILE,JOBHIST,LABDATA

WHERE EMPFILE.EMPID=JOBHIST.EMPID

      EMPFILE.EMPID=LABDATA.EMPID                   2-5

      AND LOCATION=B101AND JOBDS<08/07/86

      AND CREATNIN>0.8

ORDER BY CREATNIN
```

```
EMPID          JOBSD          CREATNIN

22222         09/02/86          0.9                 2-6
11111         05/12/80          1.2
```

Fig. 2.6b Examples of writing SQL queries.

around the database by using more than one table. We can "join" matching rows in different tables to expand on the information required. For example, we want to know something about liver enzyme levels in certain employees.

A query is constructed that will connect employees with their work assignment and creatinine results as shown in 2-3. This tells the system to match the Employee ID in EMPFILE to the same Employee ID in the Job History and Laboratory Data tables. The response to the query is shown in 2-4. It has been shown that (1) simple one-word commands can retrieve data and (2) SQL can navigate the relational database. There are numerous other SQL specifications that provide flexibility in data retrieval. The next query will find personnel in plant Location B101 whose Job Starting Date shows them working at their job assignments after 08/07/86 and whose creatinine levels are high normal or above. The query is being made in October 1986 because a solvent release in that location on 08/08/86 may have been toxic to the employees and high creatinine levels may indicate a work-related exposure. On 8/11/86 blood chemistries were run on all employees in Location B101. This query is shown in 2-5.

In response to the query, Employee ID 44444 was excluded because he did not work in Location B101, as shown in 2-6. Employee ID 33333 was excluded because his creatinine level did not meet the specifications. Note that the creatinine results are in ascending order, but they could have been requested in descending order. Constructs and specifications used in these illustrations are common to most SQLs.

Sharing a Database. The most common reaction among professional users, especially in occupational medicine, is concern over how to protect confidential medical data. This is particularly true in an integrated shared system. There are also technical problems with sharing data access. SQL has controls to resolve these problems.

Grant and Revoke. Data security in SQL is provided by the DCL. This part of SQL is available only to the database administrator. The clinic personnel, industrial hygienist, and safety manager may all be able to view data in the demographic table called EMPFILE when given GRANT permission for unrestricted viewing access. Note that access is only for viewing purposes. Writing to the file is not a privilege granted by this command.

The right to use the LABDATA table may be restricted to only the doctor and the nurse in the clinic. Thus, when the doctor or nurse logs onto the system, they identify themselves by a password that makes it possible for them to use LABDATA for either reading or writing purposes.

The hygienist may have permission to read from or write to the employee exposure file, but only the doctor may be permitted to share these data, and on a read only basis. In this case, the DBA would enter a command that grants permission for the doctor, but not the nurse, to have read access and the nurse would be unable to access the exposure file.

REVOKE does the opposite of GRANT. It removes access privileges. Certain specifications may be added to these commands that channel access even more stringently. For instance, only those employees in specified departments could be seen by a user. GRANT and REVOKE are very effective protection controls for multiuser databases.

Record Locking. On systems with several or more users, *grant* and *revoke* are only one type of system control. Another is record locking. This prevents two users in a network from writing to the same record simultaneously. Two or more users can read records at the same time, but data cannot be entered into the record concurrently. Sharing a database has many advantages, but technical problems arise as a result. Concurrency has long been an issue in database management, and there are several methods for overcoming the technical difficulties. SQL supports the record locking approach.

2.7 INTERFACING WITH OTHER SOFTWARE

Useful Occupational Health and Safety Software. Industrial hygienists and safety personnel who use computers, especially microcomputers, are familiar with spreadsheets like Lotus 1-2-3. They may use graphics packages as well. Spreadsheets are effective tools for examining numerical data. With speed and accuracy they do arithmetic operations, descriptive statistics, and some statistical analysis. They are often used for forecasting future trends in financial analysis. It is important that spreadsheets be used knowledgeably. Manipulation of numerical data can lead to invalid conclusions. Occupational health and safety measurements are not equivalent to financial data. Scientific measurements are unalterable facts. Although an industrial hygienist may find it convenient to use spreadsheet arithmetic and statistical calculations, "what if" forecasting would be dangerous in exposure assessment. However, where modeling is an appropriate objective, the ability to test a variety of possible results is beneficial.

A graphics package is useful as an adjunct to the spreadsheet. Graphing the data reviewed in the spreadsheet may facilitate curve fitting and data smoothing. Plots illustrate data and are useful in all areas of occupational health and safety. The combination of spreadsheet and graphics output provides information for managerial decision making and becomes support documentation.

Analytical statistical facilities are unavailable in spreadsheets and graphics packages, but are found in statistical software that is useful for regression analysis and specialized statistical computation.

Word processors are effective software writing tools that are used in virtually every office handling plans, programs, correspondence, and text processing. As an extension to spreadsheets and graphics, word processors provide explanatory documentation.

An Integrated Application Package. Occupational health and safety software including every kind of processing usable in the clinic, industrial hygiene, and safety would be prohibitively expensive to develop. Memory requirements would be enormous for several or more software programs running simultaneously. Combining a spreadsheet, graphics, a word processor, a database, and statistics with an occupational health and safety applications package is analogous to putting windshield wipers, a windshield washer, a gear shift, cruise control, and headlight controls all in one lever on a steering column. It can be done, but the user may wash the windshield instead of shifting into high gear. Given the present state of software technology, it is necessary to separate the applications software from software with different programming demands. Development, programming, and code maintenance will be more cost-effective and efficient. The very large, comprehensive software that would cover all occupational health and safety requirements is yet to come. The stress on integration at this time is in terms of the database.

Interfaces. An interface is a common boundary between two things. In the computer world, an interface may be either a hardware or a software link between two parts of a system. An example of a hardware link is the cable that connects the computer to the printer. A software link prepares data to be passed over the communications cable to the printer in a form that the printer can interpret.

Software interfaces are links built between applications packages so that the user can move freely from one to another. A central database furnishes a core of data around which to cluster a variety of useful software. Data subsets can be selected from the central database and placed in separate files. Interface programs pick up data from a subset file and transfer it to a specified file in another package. This type of interfacing is called *export/import*. Some transfers are made in American

Standard Code for Information Interchange (ASCII) characters, and some use software specific formats. Custom programming may be necessary to construct an import/export interface for an occupational health and safety user, but this code is reasonably easy to write and should present no serious development problems.

2.8 WHO WILL MANAGE THE SYSTEM?

A Database Administrator. In the computer world, a database system is managed by the database administrator (DBA). This title is curious because the connotation of administration is "to govern," while the database administrator has only managerial control of the database, with no control over the areas of hardware or other software. On an organization chart, the database administrator in one company might be on the same level as the managers of operations and programming. In another company, this person might report to a manager of operations. In still another company, he or she might be one level above operations and programming, reporting to the director of information services. This wide variation in authority assigned to database administrators in different companies indicates a lack of uniformity in the importance accorded the position.

A Database Manager. In the occupational health and safety environment, it might be well to have a database manager rather than an administrator. Users represent independent disciplines with the common goal of protecting human beings from workplace hazardous exposure that might lead to injury or impaired health. From the company's viewpoint, the occupational health and safety database is not on par with the accounting database. It is a shared database in which control is a managerial function rather than an executive function. It would seem apropos to call the person in charge a database manager because his or her function is to coordinate database use across the health and safety departments. All managerial functions in the use of the occupational health and safety applications software are coordinated by one individual who speaks for a user. The principal contribution of this person is to provide systems guidance to the professional users.

The Team Approach. The occupational health and safety system lends itself to team management. Because of the diversity of professional areas and technical data, it may be helpful to have a team or committee. The team could work under the day-to-day guidance of an individual called the *data manager*. There should be an emphasis on the need for guidance by the team leader, rather than management or direction, in order to promote an atmosphere of joint goals and common interests. The team would consist of representatives of occupational medicine, industrial hygiene, safety, and information services. In addition, other interested departments such as human resources, payroll, and purchasing might request representation on the committee. The decision to manage the occupational health and safety software by a database administrator, data manager, or team will depend on company politics and management policies. No single method is best for every situation.

The data manager and team should be selected in the initial stages of software development. This will permit everyone to acquire early knowledge of the computer system and of each other's interests. Joint involvement during development will engender smoother working relationships once the software is up and running. User representatives will become familiar with the advantages and limitations of the computer system.

Qualifications for Managing the Database. A manager of occupational health and safety applications software should be chosen on the basis of his or her database management skills, some of which are suggested in Figure 2.8. Excellence in all these areas will be elusive, but essential areas include user orientation, the ability to communicate well, and mediation skills. In time, other skills can be developed. In all probability, people who are equally adept with computers and occupational health and safety will not be found, and the best qualified applicant will have to be trained on the job. The pace of development will allow time for learning.

Management of the Database. The database is a user-oriented system, and there must be mutual respect and cooperation between manager and users. During and after development, differing needs and opinions will be expressed by systems-oriented personnel and users. This is where the mediation skills of the manager will be challenged. The problem will not be in defining the objectives of the occupational health and safety system, but in agreeing on the ways to achieve them.

The team approach will support effective management. There will be periodic meetings at which everyone can voice concerns in the presence of others. Each team member can benefit from other viewpoints. The manager will have an opportunity to explore with the users any possible effects of changes in procedures or the system before making decisions. Team meetings give user representatives the feeling that their viewpoints are important, but meetings may be tinged with negative reactions and competing viewpoints. A good manager can emphasize the positive areas and defuse tensions. Because the system is integrated and shared by users

DATABASE MANAGEMENT SKILLS

People Skills

User orientation

Staff management skills

Aptitude for good human communications

Ability to mediate differences of opinion

System Skills

Understanding of applications software

Ability to work with systems analyst and programmers

Knowledge of computer operations

Familiarity with technical requirements of occupational health and safety

Fig. 2.8 Necessary skills for a person managing an integrated database.

from different disciplines, it is vital that the team focus on system objectives, not territorial objectives. The manager should be the one to contact users, chair meetings, and keep users informed.

System Management. One of the requirements of a data manager is a thorough working knowledge of the system. This is the person who grants and revokes access privileges, a right that is best controlled by a single individual. Who "owns" certain data? Who has read access, and who has write access? Who can modify or delete data once they are entered? The data manager determines the appropriate hours of machine time. The minicomputer or mainframe system may be overloaded during peak hours of use, resulting in slow response time. It may be necessary to allocate hours outside of the normal workday for users. The data manager can arrange to train users on the system, monitor use of the system, and oversee the integrity of the database. Overseeing the database is vital. Data fields should not be changed without an understanding of the effect on all user information, both now and in the future. In addition to data management functions, the data manager makes certain that data are backed up regularly and that standard operating procedures are followed.

Database management systems on minicomputers and mainframes usually have a utility that compiles statistics on users. How many times was the system entered during a given period? How many additional entries were made? Modifications to existing data? Deletions? What was the total running time, CPU time, or average time for a user session? These and other statistics keep the manager informed, supporting suggestions to users for improved use of the system. This is monitoring the system, and should be a typical work practice brought to the occupational health and safety team by a person who has had previous system experience. The data manager should make periodic evaluations of the system's operation and effectiveness. Systems are never perfect, but every effort should be made to achieve maximum benefits. Evaluations by the users will also contribute to system improvement and should be regularly sought.

System Disruption. Change is so disruptive to computer systems that it has to be implemented with extreme care. A change in the occupational health and safety database, for instance, can nullify the use of data by an epidemiologist at some later date because a data field in the past had a different meaning from the present data field. Change in procedures or in the operation of the system may insert unintentional bugs that plague users. Changes can cause loss of data already in the system or nullify a whole day's (or year's) entries. Changes must be handled under the most controlled conditions.

Changes are to be expected in an occupational health and safety database. No one in the field today would be brash enough to predict what changes will be necessary in a year or two because of revised regulatory mandates, newer technology, or legal implications. This is an evolving field. The data manager should make changes only when a written change request has been submitted, circulated among all users and interested personnel, and generally agreed upon. Change procedures should be among the first directions documented for the system.

The Data Manager and the Data Dictionary. An occupational health and safety database is a centralized operation in that it is a single large database with all data focused on a single theme. The manager of this database is an expert in retrieving information from the system, just as a reference librarian in a technical library is adept at finding specialized information for readers. When the data manager works with the system from its initial development stages, this person assumes a key role in the development of the schema and data dictionary, along with the systems analyst. Data fields in the schema will be grouped and listed, and the format will be described. These lists will be reviewed numerous times to eliminate duplication, a process called *normalization*. The basic design will set the policy for future expansion or change of the database, so it has to be flexible.

The data dictionary is like the card index one uses to find a book on a given subject, or by a given title or author. When completed, the data dictionary will describe the nature and use of all the data fields in the system. It will spell out the data fields, their meaning, and their size. The data dictionary will be the permanent record of the database design, documenting all of the information pertaining to the data fields. Any changes must be carefully reviewed to ensure that database integrity is not destroyed. On a change request the data manager will want the date of the request, what the change is to be, who requests it, the reason for making the change, and the date the change is to take effect. Some of this information will be inserted into the system for historical purposes so that users will know the time periods when data have different meanings (for instance, the epidemiologist mentioned earlier in this section). Users must be helped to understand the many aspects of database changes. When changes are made to the database, the system will probably have to be shut down. Reload of the database may need to take place before normal operations can resume.

Documentation of the data dictionary is a valuable resource tool for users, especially if they are using SQL to retrieve data or a report writer to retrieve information. Users can locate names and descriptions of fields before writing SQL queries or formatting new reports.

The data dictionary can also be the basis for on-line help to users. It will probably be written by the data manager. On-line help permits users to strike a function key that erases the present screen and brings up a screen describing the format, the character of the data field, and perhaps some suggestions as to what kinds of data are pertinent for entry.

The Data Manager and the System. If a user is selected as data manager, a systems person must assist in making managerial decisions. Systems expertise is required in minicomputer and mainframe environments to make decisions like the following:

- Job priorities
- Job control procedures
- Backup procedures
- Recovery procedures
- Machine failure potential

Documented procedures and standards for system operation must be drawn up by information services personnel. Users will not be expert enough in systems to do this. Users must have written procedures and standards to ensure correct use of the system, but stated in user terms. Enforcement can be through the data manager. All documentation should be available in both systems and user areas. The data manager must ensure that documentation is updated in order to keep users current with the system.

From time to time, new peripheral devices and communications equipment will be advocated. New applications software may become available. The data manager should make certain that a full evaluation is made before adding to or changing the system, and should seek outside expertise when not enough answers are obtainable internally. Will this affect system operation? Will this affect the database? Will this affect the users?

In the event of machine failure, power failure, or other possible disasters, how will data be recovered? How frequently should backups occur in order to protect data? These are important issues; they should be understood by the users, as well as by systems people.

Team meetings are opportunities for users to become familiar with operations problems. The data manager would be wise to bring specialists to the meetings such as computer operators, communications personnel, and others. These people can explain files, expected response times, memory limitations, and storage limitations. They can acquaint users with the way in which the system can optimize data retrieval and balance unused space in the files. Informed users will be more intelligent users of the system.

Microcomputer Management vs. Management of Larger Systems. Much of the above discussion on database management may appear to pertain only to minicomputers and mainframes. However, the microcomputer environment differs only in the degree of required systems expertise. An integrated occupational health and safety system that operates over a microcomputer network also needs strong management. User representatives at team meetings should rely on a database manager to conduct meetings, keep users informed, develop policies and procedures for adoption, train users, and keep a watchful eye on potential problems such as database changes.

Documentation is also required for microcomputers. Microcomputer users, like users of larger systems, need to know what to do in case of machine or power failure. If there are ways to recover data, they need to know them.

2.9 WHAT RESOURCES ARE REQUIRED?

Commitment of Management. The most valuable resource is the commitment of senior management to acquire an automated occupational health and safety system. Unless that commitment exists, the project will totter until it becomes moribund. From the board of directors and the CEO to top management, there must be a willingness to see the project through to the end. If inhouse development takes place, the software may be two years or more in production. If software is purchased, the project may take a year to complete.

Not only must there be determination to see the project through, but there must be a commitment for financial support. This will not be an inexpensive venture, and cost estimates will surely undergo some revision before the project is completed.

Justification for the System. Top management will expect economic justification before approving a project of this size. The challenge will be to find a good method for preparing a cost analysis. A typical cost/benefit analysis will not suffice. Unlike expenditures that improve production, an occupational health and safety computer system cannot be shown to turn out more production units than before, and at less cost. The objective of the occupational health and safety system is to handle employee health and safety data in a manner that protects the employee in the workplace. Sometimes it can be argued that manual entry of data on a paper form is faster than entry on a computer, but the trick is to get data analyzed and reported without spending days trying to summarize a file drawer of paper. How does one measure improvements in employee health and safety in the workplace? Dollar values cannot be placed on health and safety, which eliminates the use of a cost/benefit analysis to justify this kind of computer system.

Cost/Benefit Analysis and Cost-Effectiveness Analysis. Brian O'Neill and A. B. Kelley (O'Neill and Kelley 1985) ably define cost/benefit analysis and cost-effectiveness analysis as they apply to human safety; their words are worth repeating here:

> *Cost/benefit analysis* measures a planned program's costs against its expected benefits, using identical monetary units of measurement—most often dollars—on both sides of the ledger. . . . (p. 3) *Cost-effectiveness analysis* compares the cost of alternative means for effectively achieving an agreed-upon goal. The means may be programs, technologies, devices, or combinations of approaches. The goals are often expressed in public policy as laws and standards. (p. 5)

While elaborating on cost/benefit analyses, the point is made that it is nearly impossible to equate human life (well-being) with dollar values because the premium our society places on human life cannot be measured monetarily. The definition of a cost/benefit analysis is something that must be measured in dollar values "on both sides of the ledger." Dollars spent on a computer system cannot be equated to human well-being. The computer does not save life or limb; it can only improve imformation handling that assists humans in making decisions affecting human health and safety.

Cost-effectiveness analysis is not expressed in monetary units. Instead, the emphasis is placed on increasing the effectiveness of performance. This can be measured by comparing alternative means for achieving a stated goal. Alternative methods for reaching the goal are evaluated, but units of measure are not necessary. In cost-effectiveness analysis, the value of human life is not addressed because that is not an issue. The issue is which of several possible methods provides the most acceptable means for achieving a good data-keeping and information-reporting system. By this type of cost analysis, top management makes a value judgment in approving or rejecting the project, and value judgments are somehow suspect in the world of business. Approval or rejection of a proposal is preferred because it is impersonal and "business-like" when made on a dollars-and-cents basis. No apologies are necessary for using a cost-effectiveness analysis rather than a cost/benefit analysis. Cost-effectiveness analysis can be very persuasive.

While ruling out the propriety of using cost/benefit analysis in justifying an occupational health and safety system, the CEO and the board of directors can still hear strong arguments in support of the request for a system. There are numerous resources already available. The new system will expand on these resources for better technical data management.

Available Resources. What does the company have on hand at present that can be used for the new computer system, and what must be purchased? Computer equipment is already on hand. Software such as operating systems and DBMSs may be usable with the new applications software. If there is an information services department, experienced staff can work with the users to develop the system. Development will include analysis, design, programming, testing, implementation, parallel operation during changeover, documentation, and training. Cost estimates will be possible for all of these items, and probably for payback periods as well.

Where company minicomputers and mainframes exist, there are probably sizable machine demands already. Careful calculations by information services will provide estimates of how much storage and memory will be required by the occupational health and safety system and will show how much additional use the present system will bear. Many companies operate at near capacity most of the time. The question will be whether expansion of the present memory, storage capacity, and CPU is more economical than adding an entirely new machine. If present capacity is strained, investigate the possibility that a company in the vicinity may be planning to purchase a larger computer and will gladly sell their original equipment at a good price. Another possibility is to consider leasing equipment. The operating staff may need to be increased as well as the equipment. Equipment servicing should also be included in cost estimates.

If additional machinery is needed, it can be described in terms of added overhead on the present system, lower response times, or too many layers of software. The more layers of software, and the more memory that is required, the more burden there will be on the traffic pathways of the system. This is analogous to traffic congestion on community streets when development exceeds arterial road capacity. Increasing the number of users forces the system to queue user processing requests. Response time diminishes with increased user demands. This trend is noticeable when the user inputs data and has to wait too long for the machine to acknowledge that the processing request has been completed.

If microcomputers will be used, how many exist already that can be deployed to occupational health and safety? Alternatively, should the company invest in up-to-date desktop hardware? Newer machines with 16-bit word sizes and 386 processors will offer faster processing. Will these be serviced in-house or elsewhere? These are questions that will be considered by information services.

Peripheral devices such as printers, tape drives, and communications hardware and software should also be considered. Are there some in storage? Will newer equipment be required? A report to the CEO and board showing a thorough examination of the costs of several possible means for obtaining hardware will supply ample data to support the project.

Review of how much required software already exists in the company and how much has to be purchased can be done in the same manner as for hardware. Operating systems already on hand should be checked for licensing permissions. Where single-user licenses exist, it may be cheaper to purchase a larger license than an entirely new package. A DBMS may be available on-site, and information services may require that this same DBMS be used for all company applications development. A policy such as this is more economical from the standpoint of maintenance and support, but the chief question is whether the applications software for occupational health and safety will run with this DBMS.

In both user areas and information services, there may be personnel who can be assigned to the project early in development. This will eliminate the problem of obtaining approval for new personnel slots. The data manager and the systems analyst will be completely involved at the earliest stages; while programmers will be needed later. Team members will use varying amounts of time as the project progresses, depending upon the kinds of decisions and input required. From the time when user testing begins, user involvement will become more time-consuming. A cost-effectiveness analysis should utilize as many present personnel as possible in order to avoid exploding salary and benefits costs. Ultimately, additional personnel may become necessary to perform data entry or to carry out routine tasks. Data entry can be very time-consuming and may overwhelm the present staff, but during development, data entry is not an urgent factor.

Another question that needs an early decision can also affect costs. Will present data be entered into the system, or will data collection begin when the system is implemented? If present data are entered, how far back will this go? One year? Five years? Entering present data can be time-consuming and expensive. There will also be minor costs for shifting employee demographic and work history data over to the new database. Training is a cost that depends upon the number of people involved and whether travel is required. Budget constraints will tend to reduce the optimal number of people to be trained if they must travel to another area. When training is on-site, training for all potential users is a luxury because the time needed away from their duties is reduced and they can begin to see the system in its natural environment. Training will include information services personnel as well as users. Vendors usually include training in the purchase price of the software, but they limit the number of participants.

Cost Analysis. The cost analysis will show costs in relation to the items listed in Figure 2.9a.

Software Development. Because software development presents unique problems, additional remarks must be made about this area of the cost analysis. The classic question is whether to purchase software or to develop it in-house. The pros and cons of the answer deserve a lot of thought. Regardless of the approach, development is a difficult, expensive, and resource-depleting task. In-house development is inadvisable unless the systems analyst is experienced in the development of this specialized application field, as demonstrated by the past history.

COST ANALYSIS FACTORS

Purchasing costs

Hardware

Software

New personnel

Costs to produce

Initial costs

Systems analysis costs

Programming costs

Testing costs

Training costs

Implementation costs

Documentation costs

Costs to operate

Day-to-day costs

Repair costs (hardware and software)

Maintenance costs (hardware and software)

Fig. 2.9a Factors to consider for a cost analysis.

Fifteen years ago, there were no commercially available occupational health and safety software packages, but the need existed nonetheless. In time, several large corporations with sophisticated in-house computer programming resources decided to develop their own specialized computer systems. All the normal preliminaries such as the cost analysis, systems analysis, and requirements analysis were carried out, and everyone was convinced of success. In every instance, projected development times were underestimated and projected costs escalated beyond reason. Once the product was completed, the obvious answer was to put it out for sale in order to recover some of the investment. There is general agreement that industry needs occupational health and safety software, but the costs of development drive purchase price beyond the reach of many businesses, limiting the market.

Central to all use of occupational health and safety software is the demand for modification to user specifications. A package acceptable to the developing corporation required modification for the purchaser. This required more expenditure and inflated the original cost of development. Clearly, this was not an economical venture. In time, even users in the developing company found it necessary to modify the software because of the evolving nature of their fields. On an in-house basis, additional costs for analysis and programming became a financial burden. There seemed no easy solution to the development problem.

Millions of dollars have been spent since the late 1970s in trying to develop occupational health and safety software. Some packages are for medical use or safety use alone. Few encompass all of medical, industrial hygiene, and safety on an integrated basis, despite the fact that an integrated approach is a worthwhile objective.

Beginning in the early 1980s and continuing to the present, individual businesses initiated development of integrated software, not only for their own medical use but also for the industrial market. The result of these efforts was financial woe of immense proportions. Dollar support during the months of designing, programming, and debugging these systems that stretched into years led to never-ending frustration and despair. At the end of the road, frustration continued with the problems of modifying the package to the buyer's requirements. Budget cycles for large corporations that can afford steep prices are on annual (or longer) timetables. From the developer's point of view, it takes an exasperating amount of time to obtain all the required signatures from information services, the budget manager, legal services, the CEO, and the board of directors, plus others, before a purchase order can be cut. Meanwhile, the financially plagued developer is twisting in the wind with a product, but no income. If it were possible to sell dozens of smaller packages to small businesses, survival between sales to large corporations might be easier, but the price has been typically out of reach for small businesses—precisely businesses that need to purchase occupational health and safety software. Very large companies have in-house specialists who can develop software if they choose, but small businesses rely solely on purchase from vendors.

What of the corporate buyer's frustrations? Users chafe at the long corporate delays in obtaining the software they so anxiously await. They find their operations disrupted during the long period between initial search for a system and final delivery and implementation. They think of more modifications that they want as time passes. In short, purchase of occupational health and safety software is no picnic for either developer or user-buyer.

It would appear from the foregoing discussion that neither in-house development nor commercial purchase is a viable possibility. Seen in the above historical context, this comes close to the truth. The real issue has been the lack of computer technology to overcome the problems of development, programming, and modifications. With the newer technology of 4GL tools described in Section 2.4, however, there should be a breakthrough in the deadlocked development of integrated software for occupational health and safety.

For the developers, screen painters will eliminate untold hours of programming and permit quick modifications to screens and the data dictionary. Report painters will encourage more flexible report formatting; previously, every report format had to be programmed. Programming bugs will be reduced. All of this will reduce the strain on developers' pocketbooks. Even so, there will still be a need for screen and report painters that are specially designed for occupational health and safety and can provide features critical to the field—for instance, table look-ups that help to avoid complete entry of the same data over and over again. If a name is entered on a record, it should not be necessary to enter the social security number and age as well. This can be automatically looked up by the system and returned to the record. With 4GL tools, the cost of developing occupational health and safety software will be brought under control. A buyer should purchase both the applications software and the screen and report painters. This permits future modifications to be made by the purchaser and saves a lot of money. Conversion to microcomputers will put all of this software into small plants if prices can be brought into the range of the small business.

The preceding paragraphs emphasize the need for caution in deciding whether to attempt in-house development or to purchase a package. Cost comparisons may be valid illustrations of cost-effectiveness in software purchase. Modification without a screen painter will take time and money, however, and the risk of bugs will still be high.

One other option is worth researching for cost comparison. This is an on-line service bureau by a company equipped with occupational health and safety software. This may also be useful on a temporary basis until the purchased system is up and running. It offers an additional possibility for entering present data without hiring temporary help to accomplish the task.

Planning and Evaluating. Cost analysis should include all the steps in the development process. For in-house development a systems analyst will probably already be available. This person will confer with the users at length to obtain a thorough understanding of their needs. It may be worthwhile to compare the cost of the systems analyst from information services with the cost of a systems analyst consultant. Hands down, the in-house person will win the job because the consultant may charge on an hourly basis, and this is a time-consuming assignment.

The consumption of time in the development of the database will be another expensive factor. The database must be designed, and carefully constructed and a data dictionary made. As the design progresses, a complete list of tables must be kept to link data entry with related information. Screen design takes time mostly because of numerous user revisions. A screen painter will slash costs for screen development. It will also help to develop the data dictionary. The systems analyst will design navigation paths to facilitate user movement through the system.

Once coding begins, the systems analyst will work closely with the programmers to ensure that the system follows the design and that the design works. No system ever follows the original design explicitly. As programming and testing take place, the need for modifications continues. As implementation progresses, the systems analyst will gradually shift oversight to the data manager, who has been involved since the early stages of design and has worked closely with both users and the analyst. Thus, time costs will cover the systems analyst, data manager, user team, and programmers. During testing, include time for programmers, users, and others. Add on management time for reviewing the project with the systems analyst, the data manager, and others. No single method of development will offer a cheap solution, but some methods may be better adapted to corporate policies and procedures than others. Use these in the cost-effectiveness analysis.

Once planned and implemented, the system will require and repair of the code. Supplies will be needed, as well as projected additions to the system, some in hardware and some in software. What kinds of changes to the system can be expected, and what ballpark figures can be projected for them?

Benefits. To balance out the cost-effectiveness of the proposed system, there must be a strong emphasis on the benefits to be achieved. Some suggestions for thought are listed in Figure 2.9b. All of these possibilities may not be present in every system design, but the list will provide some ideas on what may be most desirable. The list could be prioritized to show those features that can be phased in at a later date. Phasing in is one way to reduce immediate costs and to spread costs over a period of time. The original design should contain as much of the total project as possible in order to accommodate all user needs.

SUGGESTED COST BENEFITS OF A HEALTH AND SAFETY COMPUTER SYSTEM

Readily available information
Simplified data collection
Elimination of manual searches of paper files
Improved scheduling of surveillance
Improved scheduling of safety training
Compliance support
Legal support
Fast retrieval for employee and community right-to-know
Simplified inventorying of hazardous agents
Simplified tracking of employee exposures
Simplified retrieval of MSDSs
Standardization of confidential data
Identification of patterns of injury and illness
Tracking of medical care costs
Tracking of employee absenteeism causes
Flexibility for future safety and health data needs

Fig. 2.9b Suggestions for the benefits of an integrated occupational health and safety computer system.

2.10 SYSTEMS ANALYSIS AND DESIGN

Analysis and Design. Analysis means looking at a problem critically and evaluating all its parts. Once evaluated, ideas for future improvement are offered as part of a design. Designs can be textually described, but they are best given in drawings or models. In Section 1.1 an analogy was drawn between the systems analyst who analyzes and designs computer systems and the architect who designs houses. An architect consults with the person(s) who want to build the house, renders the proposed house in a sketch, and draws blueprints for every aspect of the structure that the builder must follow, step by step. The rendering and blueprints are designs for the proposed house. By analogy, the systems analyst draws two-dimensional models of every aspect of data flow in the proposed computer system after extensive discussions with users.

The Systems Analyst. An essay on the important function of the systems analyst as a salesperson is a humorous piece by William E. Perry of the Quality Assurance Institute (Perry 1987). Perry buys a car after a flawless sales performance. When he later takes it for repair, there is no salesperson, and he becomes frustrated by the repair mechanics, who appear to have no personal sensitivity to him as car owner. He compares the salesman with a systems analyst who does an excellent sales job and then leaves the computer user to fend for himself in dealing with the computer technicians when repair is necessary. The automobile salesman and the systems analyst "please the customer and sell the product" superbly well. "Most technical people do not care about their customer . . . even though it is logical and would make us, the customers, happy," laments Perry. The comparison has a lot of truth—not because the technician is heartless, but because it is not his function to deal with the user. A suggestion that follows from Perry's complaint is that the system analyst's role should extend beyond the initial purchase to the repair or maintenance of the system, when a user-oriented buffer betweeen user and technician is helpful.

In *Managing Management Information Systems* Ein-Dor and Segev (1978) claim that a systems analyst must have "logical ability, thoroughness, ability to work with others, resourcefulness, imagination, oral ability, abstract reasoning, emotional balance, interest in analysis, writing ability, curiosity, decisiveness, empathy, mature judgment, practicality, ability to observe, dislike of inefficiency, initiative, integrity, intelligence, interest in science and technology, interest in staff work, numerical ability, open-mindedness, and selling ability." Needless to say, this is a lot to expect of one human being.

What these quotations illustrate is a user orientation that enables the systems analyst to communicate well, identify with the user, and translate the user's views into computer technology. The systems analyst is the user's advocate. In the end, the analyst "sells" solutions to both the user and information services.

Information Services. Information services is service oriented, as its name implies, the services ranging from user contact by the systems analyst to code writing by programmers, and from machine operation by computer operators to machine repair by computer service people. The systems analyst has a working relationship with all of these people, knowing whom to contact for any job, and with a sensitivity for the constraints of each job.

Information services departments are organized around the effective operation of computer systems. The common goal is to manage data processing for the company in order to produce timely information for decision making through reports and on-line data for all levels of management for effective communication throughout the company. The company depends on information services to maintain an uninterrupted flow of information, which is no minor task. If information services sometimes seems demanding to users, it is because altered or modified data can distort correct information, time lapses can cost money, and poor control of company data undermines company stability.

When a user requests service, the channel is the systems analyst, who may be the continuing contact for the user. The analyst forges a path through the labyrinth of computerland. The systems analyst makes personnel such as programmers and operators anonymous participants in user service and protects the programming staff from direct user contact. A partner to both users and colleagues, the analyst is often in an uncomfortable position when complaints are voiced on either side, but the analyst is a diplomat and knows how to walk a fine line.

Tools for Analysis. Upon a request for a system, the analyst assigned to the case will visit the users. For an integrated system, users in all the areas will undergo the same kind of evaluation. During this and subsequent on-site visits, the analyst will learn the reasons for wanting to implement a computer system in occupational health and safety, take note of the departmental layout, look at available computer equipment and software, ascertain the working interactions between users, talk to other personnel, check available space, and collect copies of all printed forms in present use. Any personnel involved with present or future data management will be interviewed at length. Sometimes questionnaires will be used, but analysts tend to rely more on conversation and direct observations for their information.

These talks provide insight into present problems, including personal attitudes. At first thought, the latter information would appear to be intrusive, but attitudes toward a new computer system can frequently spell success or failure. Personality conflicts enter into the distribution of assignments when the new system arrives. For example, the department manager may be completely convinced that a computer system is necessary, but the manager's top aide may see the system as frightening, or as a competitor for the manager's attention. The manager turns responsibility for system design approval over to the aide. At each level of design and implementation, the aide has nothing but complaints to report back to the manager, and the analyst soon recognizes that little can be done to overcome the adversarial relationship. Personality conflicts within the department can have the same deleterious effect on system development. A skillful analyst will spot these problems early and work tactfully around them. Sometimes it helps to interview off-site—for instance,in the analyst's office—in order to give a person more opportunity to speak candidly.

Collection of paper forms provides the analyst with a library of data important to the user and normally kept on file. This is an invaluable source of information and appears to users to be the foremost function of the analyst. The analyst will want to know which data need to be modified or added. If data are collected in text format, can the data be represented in brief, or coded, entry? The paper forms are the beginning of computer screens, giving the user the first glimpse of a real system.

Any procedures manuals will be carefully read, and work practices noted by the analyst will be meticulously recorded. Medical personnel and some industrial hygienists are good at maintaining procedures manuals in their departments. Where procedures manuals do not exist, extensive observations of work practices will be made.

Other tools for analysis may be employed by the analyst, but those chiefly relied upon are interviews, observations, collection of forms, procedures manuals, work practices, and possibly questionnaires.

Tools for Design. Probably the first designs the analyst will draw up are organizational charts showing the people who will be involved with the system.

Along with these will be flowcharts to show where data come from, how they are processed in the system, where they go, and how they are reported out of the system. This is an extensive modeling task. Numerous types of structured models are employed in systems analysis, and analysts generally have preferences for certain models. Sometimes there is a standard company model to which the analyst must adhere. There are also software packages that systems analysts may use for graphical modeling.

After flowcharts and models, there will be charts showing database elements and their relationships. Along with recommendations for system development and implementation, there will be scheduling charts, possibly of several types. All of these tools will accompany a formal proposal for the system. Users, information services, and others will make changes in the initial proposal before final approval is sought, and it will be the systems analyst who pulls it all together.

Prototyping. The trend in systems analysis is to favor prototyping over writing detailed specifications for system development. Prototyping is an important tool for the analyst. Its chief advantage over specifications is the iterative development process that accommodates the evolving nature of system development. When specifications are written and shown to the user, the user interprets them according to his or her own frame of reference, which may not be the manner in which they are understood by the analyst or programmers. Coding is completed and the user sees the product, only to be shocked that things are not as they were expected to be. Much time, energy, and money have been expended to reach this point, only to have an unhappy user who wants to revise extensively, involving more time, energy, and money.

A prototyped system design will be more a product of occupational health and safety professionals than of systems professionals. The area where this will be most noticeable is in screen design. Users do not review code; they view screen designs on the VDT or on paper, basing their approval or rejection only on what they see and how the data are entered or retrieved. The system does not have to be in an operating mode. If they like what they see on paper or the VDT, they look favorably upon the system.

If a screen painter is used, screens can be constructed speedily, shown to the user for review, and quickly changed to satisfy the user. In fact, in the eyes of the prospective user, a screen painter produces such convincing evidence of an acceptable system that the viewer fails to realize that a fully working system does not yet exist. It is tantamount to approving a car that looks terrific but lacks running parts under the hood.

Development is an iterative process. Design, show it to users, receive comments, redesign, show it again, and receive more comments. This process is repeated over and over again until final approval is obtained. If screens are programmed, code is rewritten over and over, consuming a lot of hours and available funds. Costs expand exponentially. The most functional and cost-containing control over system development is to reduce coding by reviewing screen design first on paper or a VDT. From duplication of the original paper form to the final screen form, text entry can be refined by concise codes, screens can be combined or separated, and data and screens can be functionally grouped for easier handling. Suggestions will come from both users and the analyst. The final outcome of this iterative process will be the gradual transformation and refinement of a system in which the user can have pride, having participated in its development.

At completion, the users see a system they are familiar and comfortable with. The systems analyst has had ample opportunity to observe and record user difficulties, and these can be addressed in the documentation that benefits future users. As the screens are designed, the data dictionary can be developed, whether a screen painter or a paper design has been employed. Prototyping is the analyst's most powerful cost-effectiveness tool.

2.11 WHAT DOES THE DEVELOPER CONTRIBUTE?

A Partnership. The primary contribution of the developer is service to the user. The developer provides the computer mastery needed to produce the occupational health and safety system. The developer also advises users by pointing out options, offering suggestions for improved system operation, and assisting in the continuing process of system maintenance. A developer-user relationship becomes a partnership for the management of vital technical data.

Partnership is a compelling necessity. Users are best fitted to determine the priorities because they have the expertise in occupational health and safety, and a developer who focuses on a service role will remain sensitive to user opinions at all times. Forcefulness on the part of a developer can cause users to shelve a system from frustration and dissatisfaction. A system may be an electronic marvel, but an unhappy user will consider it worthless.

In addition to forging this positive relationship, the developer should aim for a system that is as cost-effective as possible and be committed to user support and open communication. It is certain that the user will require further development as time progresses.

The stages that the project will pass through are:

- Analysis
- Design
- Development
- Implementation
- Evaluation
- Acceptance

The Developer's Contributions to Analysis. The developer contributes to analysis by helping the user to organize thoughts, define problems, and devise solutions. Problems with recordkeeping are usually multiple and multipartite. Single solutions are unlikely, but multiple solutions can get confusing. To reduce confusion and enhance decision making, feasibility studies and requirements studies are done. The feasibility study determines whether something *can be done*, while the requirements study describes something *to be done*.

A feasibility study will show users whether an integrated occupational health and safety computer system can succeed. A few of the options to be weighed may be:

- Which hardware selections will work, and how well?
- Which software is needed?
- What staff adjustments or additions are needed?
- What are the constraints in each area?
- What are the costs? (Provide ballpark figures only, if these are all that are at hand.)
- What company policies must be considered?
- What procedures must be followed?

Based on these preliminary decisions, users can set their priorities. This will delimit the mission of the project. A requirements analysis should follow, proposing the future path and initiating design of the system. The developer has begun a user education process that will create discerning users.

As the developer studies user data, work practices, and reports, work flow can be illustrated for the user. Two-dimensional models help users to see themselves in the context of recordkeeping and the flow of information through the occupational health and safety departments. Normally, we measure our responsibilities by the nature of our work assignments, the forms we fill out, the records we keep on file, and the reports we make to someone else. The developer will bring fresh insights to these processes, insights that will suggest areas where our work performance can become more efficient and effective through the use of a computer system.

Contributions to Design. "The design team should be user driven". (Gould and Lewis 1985). They urge developers to adopt three principles in the user-developer design relationship:

- An early focus on users and tasks

- Simulations and prototypes for observing, recording, and analyzing user reactions
- Iterative design

A goal in system design will be to minimize user frustration, provide easy data input and retrieval, and make system navigation compatible with user work practices. The system should establish user confidence by ensuring that access to the system will be controlled, data will be kept confidential, and security will be provided against data distortion.

Design is tiring because of the tedious detail. Insufficient detail will hamper programming and plague the system with acute and chronic ailments. The designer juggles these details expertly. The designer is a performer bent on giving the customers what they pay for.

Modeling is a communication technique used by the systems analyst to explain to the user how the system will work and to explain to the programmers how to code the system. Models become system specifications that programmers examine for codability. Prototypes germinate in this design stage. Menu designs, screen form designs, and report designs follow system models. Database design is especially tedious, with numerous structured diagrams needed to define files and show relationships among files. It takes a lot of paper to illustrate an entire system, but designs clearly communicate the intricacies of the system to users, who should examine the specifications for workability in their environment.

Designs become part of system documentation. In this manner, the models have significance for future repair and maintenance of the software. Detailed design of the system is critical to effective maintenance, which is why it is worthwhile to take the time and make the effort.

Contributions to Development. Coding and system testing are development activities. At this stage, the user is on hold while the developer gets the system up and running. Once the occupational health and safety software is complete, it is expected to be operational for many years. It is the developer's responsibility, therefore, to code something that can stand the test of time.

Programmer-written code for an integrated system will comprise thousands of lines. Reliability must be built in from the first to the last line in this massive, time-consuming effort. In time, change will be inevitable. Changes in a smoothly operating system increase the risk of injecting unreliability into the software. Company policies and procedures also change under the influence of changing internal and external forces, some of which are shown in Figure 2.11.

Structured programming is a powerful aid to code reliability. In the

INFLUENCES ON POLICIES AND PROCEDURES

Federal, state, and local government statutes
Legal viewpoints reflecting latest court actions
Work practice changes
Personnel shifts
Hardware and software add-ons and upgrades
Special influences (unions, community groups, etc.)
Changing professional practices

Fig. 2.11 Internal and external influences on company policies and procedures that will affect the occupational health and safety computer system.

design stage, the project is broken up into modules. Each module is devoted to performing one task. As the task is defined, characteristic functions become apparent. The objective of the modular method is to isolate functions in small programs that fit neatly into appropriate places within the larger design. Small sections are programmed individually and tested. Then, they are joined together into larger sections and retested. By putting together a few sections at a time until a module is completed, the program eventually grows to its full size. Like a house under construction, system building takes place in carefully planned stages.

If there is strict adherence to good coding standards such as comment lines, indentations, and other programmer work practices, the modular approach makes it easy to take a coded program apart for further work. For a new programmer coming on the job, the style of the code will be difficult to emulate, and modules and functions have to be of manageable size. For a new programmer also, it is imperative that documentation of the original code be available. Documentation and code have to be written so that any programmer can follow them. Creative programming has its place, but not in a complex system that requires long-term maintenance. Strict programming standards need to be adopted and sustained.

The foregoing remarks pertain to coding by programmers. The faster, more efficient way to code applications software is with screen and report painters. This is also the easiest and fastest method for prototyping. The dozens of lines of code behind a screen design are automatically generated by the computer itself. Code debugging is unnecessary because the system generates easily compilable code. Within hours, it is possible to display a screen for users to critique. Only minutes may be necessary to modify the composed screen according to users' comments.

System testing is a developer's domain. Testing is the time when the system is put into action to determine that the processing of data produces reliable and accurate results. Bugs, the programming glitches that exist no matter how much care is taken in writing software, are sought and fixed. These may be the result of poor logic or incorrect language syntax. Planned, systematic testing should be conducted. Testing with imaginative data dreamed up by programmers is not advisable. Testing with real data should include extremes as well as normal limits in order to verify the reliability of the system. Test data do not have to be live, but they must be realistic, not imaginative.

There should be testing for systems criteria. Is user access controlled? Are the data validated at entry? What is the machine response time? Are the files protected? Are there controls that prevent the user from accidentally deleting records? Are the bugs ones of syntax (incorrectly stated code) or logic (incorrect reasoning or a faulty algorithm)? The developer should try to force system failure.

When testing is complete, bugs have been removed, and retesting has ensured a working system, the users can be trained. This is a transition period during which users in training will note final corrections or modifications to be made. Developers never have enough time to do all the things users would like them to do. Particularly during training. This forces users to learn to depend on themselves after a reasonable period of time.

Contributions to Implementation. Implementation is a time of user testing and evaluation. Only exceptional development and minor miracles will preclude the possibility of finding additional bugs after system implementation. However, instead of checking for bugs and content, the chief questions are:

- Does the system perform as the users expected?
- Can data be entered and retrieved efficiently by users on a daily basis?

The developer will be looking anxiously for the answers. Some further adjustments may be necessary to ensure the right fit of software to users.

The developer's first step in system installation will be to put the system on the hardware and get it running. Systems personnel will be the first persons trained; and training for users will follow. Afterward, users can resort to hot lines and documentation for further assistance.

Evaluation and Acceptance. These are chiefly user activities, although the developer will be available for consultation during evaluation. Users will need answers to questions about operation, possible improvements, and modification or repair. Evaluation and acceptance are discussed more fully in Chapter 5.

2.12 WHAT DOES THE USER CONTRIBUTE?

User Contributions. Users include professionals, semiprofessionals, and nonprofessionals. The occupational physician may use the system to check out some information, but a nurse will be likely to use the system on a daily basis for entry and retrieval. A lab technician may enter blood chemistry results. The clinic receptionist will set up examination schedules and may enter other examination results.

In the industrial hygiene department, the hygienist may be the only person to actually use the system, but it is possible that a secretary will enter some data. The safety manager will probably delegate data entry to a secretary while using the system for retrieval of accident data and analysis. Both the hygienist and the safety manager may want to schedule surveys and inspections themselves.

The occupational health and safety system is not designed for traditional clerical use; it is a technical data management system, not a word processor. Two classes of users will contribute to its design: the professionals and their clerical assistants, each group having different points of view. A list of probable users of the occupational health and safety system is given in Figure 2.12a. In a large corporation, there may also be special people assigned to enter data for MSDSs, perhaps from the company library.

The contributions of these people will span a broad range of system requirements, and their typical specifications for the system will be like those shown in Figure 2.12b. In brief, they will expect the systems analyst to answer all their prayers, but they will soon discover that the analyst is not omnipotent and that adjustments must be made because relative terms like "fast" and "easy" mean different things to different people. And "cinch" does not exist equally for all users.

TWO CLASSES OF USERS

Professionals

Physicians

Nurses

Medical technicians

Safety managers and directors

Safety supervisors

Industrial hygienists

Industrial hygiene technicians

Clericals

Department secretaries

Data entry operators

Fig. 2.12a The users of an integrated occupational health and safety computer system.

A Needs Assessment. Users consider computerizing for certain reasons: they are dissatisfied with manual methods, they've heard that computers can improve office operations, or they have too much to do and not enough personnel to get it done. Before a system is designed, the analyst will need to know what the users have in mind. This inquiry focuses on problems that need to be resolved by installing a computer system. It is possible that user problems cannot be totally solved with a computer. The needs assessment is a statement that gives the analyst a place to begin; it is the user's initial assignment, and it will become the basis for the system blueprint that the analyst draws up. Once the analyst has assesses the needs of the users, and what the users consider to be their goals, the feasibility and costs for developing such a system can be judged.

Needs generally arise from irritations, gripes, or poor function and quickly evolve into wish lists. A needs list should not include solutions; solutions are left to the analyst. The most helpful way to develop a needs list is to state both the need and the reason for the need. The list might look like the one shown on page 98.

Need	Reason
Better way to control work load	Paper volume is out of hand
	Have to hire more help
Faster way to write reports	Takes weeks to organize data
	Can't find data quickly
	Ties up time needed to do other things
Cost-effective office procedures	Time is money
	Budget won't allow more help
Keep up with work demands	Getting further behind
	No time to deal with problems
	Bottlenecks with paperwork
Cope with new responsibilities	Keep getting more to do
	New requirements take time to learn and implement
	Can't handle more records
More time to write records	Records are done manually
	Neglecting to get some records completed

User Design Contributions. Let the user beware! With every seemingly simple request, there will be trade-offs. Some examples:

1. Instant retrieval depends on how efficiently the data are stored. If a whole paragraph is used to describe one of the causes of accidents,

TYPICAL USER SPECIFICATIONS

Simple English directions

Technically accurate data

Instant retrieval

Easy data entry

Fast learning curve

Flexibility and easy modification

Tight security

Valid data

Guaranteed reliability

A cinch to operate

Fig. 2.12b Specifications that users require in an occupational health and safety computer system.

instead of one or two key words, it may take the system longer to search for the desired information.

2. If field prompts are in technical language familiar to the professionals, rather than in English familiar to the data entry operator, the operator will need a long time to learn this complex system.
3. If doctors and nurses become impatient with the number of moments it takes to save their data and proceed to the next screen form, they may have to forego some data validation. Repeating each page of entry to ensure correctness of data is excessively time-consuming, but checking validity ranges and data types may be a compromise.

With careful guidance by the analyst, users will plan requirements pertinent to the system. In balancing trade-offs, users learn the limitations of the system and adjust their expectations to realistic levels. This process helps to forge the necessary partnership. The analyst becomes acquainted with users' personalities, and the users learn to trust and depend on the analyst for guidance.

Purpose of the System. The purpose of the project is defined by the users at the beginning of analysis. It should be etched in stone and read frequently by users to keep the project from diverging. Project definition poses problems for users because their objectives differ according to their individual responsibilities, as exemplified in the above wish list. The purpose of the system tends to be based on prevalent problems as users perceive them, and this presents a real problem. The perceived problems may not always be the real problems. User-described problems are usually manifestations of an underlying problem or set of problems.

To illustrate this, consider the following complaint:

> It took three eight-hour days of going through three file drawers for me to locate the data that the risk manager needed before we could write the report. If this had been a computer system, I could have had the information in less than an hour and saved twenty-three hours to spend on my more pressing duties.

When the analyst looked into the problem to see what a computer system could offer, he found that it was not the search of the three file drawers of folders that took so long. The real culprit was the manner in which the data were filed, and there were three underlying problems, not one. The first was that the forms provided 10 lines of space for written text to describe the observed condition. Each entry had to be read for content before the condition was categorized according to the risk manager's criteria. The second problem was that there were four different examiners, and each had unique terminology for what was apparently the same condition. The complainant had to make hindsight judgments about the equivalence of their assessments. The third problem was that this was company information, and the risk manager's description of these conditions differed from this manager's description. Consequently, the conditions were categorized differently in their respective departments. The analyst's conclusion was that a computer system would not solve the complainant's problem unless the real issue was faced: The form for capturing data would have to be revised for efficient data retrieval and be compatible with data capture in other sections of the company.

The identity of the complainant was deliberately omitted. The reason is that this problem is so typical of occupational health and safety data management that the complainant could have been the physician analyzing a health condition, the industrial hygienist analyzing the prevalence of an agent exposure, or the safety manager analyzing the cause of certain accidents.

There will be many reasons for instituting an integrated occupational health and safety system. The complainant's story illustrates the necessity of separating the symptoms of a problem from the actual problem. A set of problems and a set of manifestations for each may exist. To establish objectives for the system, prioritize them so that there are primary, secondary, or even tertiary problems. Objectives can be stated in response to finding solutions for the problems. For instance, the primary objective of the occupational health and safety system may be to improve oversight of employee health and safety in the company workplace. A secondary objective may be to readily identify causes of accidents. A tertiary objective may be to control expenditures for workman's compensation and disability. In some companies, this tertiary objective may be the primary objective and the primary objective may be the secondary objective. Priority ratings will reflect the company's philosophy or policy.

The complainant's problem is also associated with a serious misconception about computer systems: that electronic files will do away with paper files. It is true that paper files are space hogs. They accumulate the dusts of time, but no one dares throw them out because legal evidence may be needed, the government mandates that they be kept, or there is a hoarding tendency in the company. The misconception is that all the data can be put into electronic files. Computer users soon discover an inescapable truth: Computers generate paper, and lots of it. No one wants a computer without a printer, and printers output *paper!* Paper files will still be a necessity. (See Section 4.4.)

Forms. The most direct method of helping the analyst begin the analysis is to collect every form a user employs in the conduct of business. These forms reveal all the routine data kept on file, make requests for action (requisitions), and transmit information to others (reports). There are also report-like form letters that transmit results of examinations, laboratory tests, and surveys. There are calendar-like forms used for scheduling medical exams, sampling events, and meetings. If the forms are multiple, the analyst will want to know the number of copies and their distribution. The analyst will also want to know the purpose of each form and the occasion for its use. No form should be dismissed as immaterial. Who fills out the form? Who signs it? Who approves it? How many hands does it pass through, and in what order? Design begins at the forms investigation stage. Joint study of the forms by the analyst and users may reveal the potential for discarding some data (or an entire form) and combining other data.

Work Practices. Every user has a daily routine to follow and some less typical, special-purpose routines. The analyst will note each routine and its frequency, particularly where data capture occurs. As the analyst goes over the forms and routines, users commonly wish that they could add certain pieces of information or simplify something. This is the time to tell the analyst. Detailed scrutiny of the forms at this time evokes improvements.

The forms will illustrate links between people, different related department routines, and how one department relates to another. These links can be matched by programming into the system special jumps from one type of information to another. (See Section 3.8.)

Work practices help determine how to group the screens into units (programs) and how to group units into modules. Grouping will save time. A data entry operator laboriously entering medical examination data will be very frustrated if he or she has to back out of one group of screens to reach another group of screens in order to enter a different kind of test data. Backing out means that several or more menus (selections) are entered and the pathway is reversed to reach the point of divergence, only to choose the next pathway. Common pathways can be designed into the system when defined at early stages. Menus and jumps are methods for making software flow with normal work practices.

Filing Procedures. How are the records filed in each department? By employee last name? By employee ID? By employee social security number? Are agents filed by common name? By Chemical Abstracts Society (CAS) number? By agent ID? Do all departments adhere to one filing policy, employee ID, for instance? Perhaps the departments differ and the clinic files examination records by employees' last names while the safety department files accident reports by social security number. If departments differ, the analyst may include cross references such as table lookups in the system. (See Section 3.7.)

Mathematical and Statistical Routines. While going through data on the forms, the analyst should be informed of any transformations, calculations, or statistical routines that are performed on numerical data. An example would be spirometry testing in the clinic. Are raw results entered? Are calculations made that include machine temperature or allowances for sex, age, and race? Are predicted values read from a chart? Is the spirometer one that outputs results from its own computer with all adjustments completed? If the answer to this last question is yes, how will the spirometer results get into the occupational health and safety system—by manual entry or by an interface between the spirometer and the larger system?

Using another example, does the industrial hygienist anticipate using the new system to select certain exposure values and run them through descriptive statistics tests giving ranges, means, standard deviations, and standard errors? The analyst will want to give careful consideration to how the system can aid the hygienist's selection procedures and the calculations that will follow.

Constraints. Once all the forms are analyzed, along with the work practices, company policies, and user hopes, design can begin. The user's ideal system can be considered, and what it will take to achieve this. However, there are always constraints that temper these wish lists.

The company's legal department or risk management will express an interest in the occupational health and safety system. The legal department will surely want to be aware of recordkeeping changes because they may have to go to court with sensitive data. The system will need to follow any recordkeeping requirements laid down by OSHA, EPA, and state and local governments.

Budget constraints are so commonplace that it almost seems trivial to mention them, but they need to be considered in terms of their ultimate effects on the new computer system. Will new personnel be required? How much upkeep will affect the departmental budgets? Will the new system reduce the hygienist's recordkeeping time enough so that a second hygienist will be unnecessary? The answer to the last question is probably no, but if the hygienist has this item on his or her wish list, it needs to be addressed. This brings up a salient feature of computer systems. Once a system is installed and mastered, the user invariably thinks of new ways to use this electronic tool to further advantage. In time, more ideas are spawned, and the computer system continues to evolve and expand. As expansion occurs, departments follow. Bear in mind, however, the caveat to this prediction: Expansion follows user familiarity with the system, and this depends upon whether the physician, industrial hygienist, and/or safety manager gets deeply involved in computer use.

The Requirements Analysis. Requirements for the ideal system will follow the analyst's investigation. When systems personnel use the term requirements analysis, users tend to feel obliged to accept these as *the* specifications. This is not true. This is a time to rethink the project. Were sights set too high? Were things overlooked that might have been included? Can the system be supported in the future with regard to both budget and systems maintenance? Can present personnel handle the new system? Does the proposed plan stand a chance of being approved by top management? If not, what revisions will make it more palatable? Users should feel free to revise the plan. Asking top management to approve the plan means that the future credibility of the department heads may be on the line. The political implications cannot be ignored.

Screen Design. Assuming that the initial investigation and plan have been successfully completed, the next level of design will be further forms analysis for data content and structure. Along with development of the data dictionary will be design of the screens. Users are the judges, and they should study screen designs to determine comfort level and workability of the screens. If a screen painter is used, prototype screens can be evaluated by users, save a lot of time, and give users the pleasure of participating directly in system design.

Report Design. Report design begins with a review of existing report forms. One method of system design is to design reports first since required output will dictate the necessary input. In the occupational health and safety system, the emphasis is on managing a large variety of data for efficient retrieval and review. The number of required reports is small, which is why the chief emphasis is on designing the database and getting the data into the appropriate input form.

Some reports like profiles will have to be specially programmed; others can be designed and produced by the report painter. Reports, like screens, should be prototyped. Reports are discussed in detail in Chapter 5.

Training. Training of users should never be done by systems personnel such as programmers and operators. These specialists are unable to adopt the user's viewpoint, their minds being attuned to machine logic and machine behavior. Users relate to patient explanations of user-oriented teachers who provide flexible approaches that suit the differing needs of learners. Many users cannot understand a manual describing the system in its own jargon. The jargon first has to be translated into nontechnical language and then learned. Fear of doing something to the system has to be overcome, and the users have to gain confidence in themselves. Once past the initial hurdles, practice can begin. Some call this the "training wheels" approach, a very apt description.

When the training wheels are removed, real use of the system begins. Logging on and off the system, menu use, and simple data entry will follow. Hands-on mastery of the basic operations leads to progressively more difficult tasks. Time-out for doing things without direct supervision is helpful. This process is time-consuming, which is the point; one should avoid crowding user training into the shortest possible time schedule.

Users are most likely to learn when training time is totally separated from work schedules. They expect trainers to be at their best; they should expect no less of themselves.

User Testing. Users who keep detailed notes at the time of testing will provide the analyst and programmers with the tools they need to debug and polish the system. User testing differs from testing by the developers. Developers will never be able to challenge the system the way a user will because they know its capabilities and limitations. They instinctively avoid the pitfalls. Unsuspecting users do not challenge the system intentionally; they bumble through the system, which is good. The system needs to be challenged for its tolerance limits and potential failures. Users make a valuable contribution to the final success of the system through this kind of rigorous testing.

Users contribute to system success by supplying real test data. An example focuses on audiometric exam in the medical module. Enter raw data from a test. These do not have to be data from actual records, but they should be typical data. If corrections for presbycusis are expected to be furnished by the system, are they correctly calculated? If baseline readings should be shown on the screen with the current readings, have they been brought back correctly from the records? Are calculations for shifts and/or significant threshold shifts (STS) correct? If the employee's age at the time of the test is calculated by the system, is it correct?

The industrial hygienist and the safety manager need to check out the same factors: the accuracy of retrieved data, the validity of data acceptance, and the results of system calculations.

Evaluation and Acceptance. Users and developers together need to establish early what constitutes fair evaluation of system success. Criteria can be defined that meet with the approval of both the systems analyst and the users. User testing should include procedures to evaluate the criteria with written conclusions at the end of user testing.

2.13 SUMMARY

The primary users of the system are physicians and nurses, industrial hygienists, and safety personnel. Secondary users may be risk managers, epidemiologists, and governments. The system is adaptable to mainframes, minicomputers, and microcomputers, or combinations of these. The underlying operating system and DBMS support the occupational health and safety application software. Specialists from the computer systems field and the user application fields are all needed to guide development of the occupational health and safety system.

Since software development is labor intensive, it is desirable to employ tools that automate programming, like screen painters and report generators which are classified as 4GL tools. The 5GL tools, like expert systems, are still in their early stages but hold promise for the future as decision support tools.

The integrated database is a collection of facts and figures that keeps the company informed in several or more related areas. A database is designed to avoid redundant data; it can be based on hierarchical, network, or relational models. It is important to build the applications software over a DBMS that will coordinate data handling, the data dictionary, and querying of the database. Database querying is an ad hoc procedure for on-line access and data retrieval when a relational database exists.

An interface is a common boundary between two kinds of hardware, two kinds of software, or hardware and software. Data can be imported into the occupational health and safety database from personnel amd payroll records and exported to statistical packages through interfaces.

In operation, the occupational health and safety computer system will require a database administrator or manager who works closely with a team of users and operating personnel. This person must have both interpersonal skills and system skills. He or she will control the data dictionary, which describes the nature and use of all data fields in the system.

Company resources for development of the occupational health and safety system will require the commitment of senior management and a cost-effectiveness analysis because the value of the system to human safety cannot be expressed in monetary units.

Development can be done in-house, by contract, or by purchase of a commercial product. An occupational health and safety system is very expensive to produce. Development begins with systems analysis and design. Analysis begins with user forms and interviews to determine what exists and what is needed. Tools for design are charts and models which can be automated by computer-aided software engineering (CASE) tools. Generating screens with a screen painter makes it possible to produce rapid, reiterative prototypes for review and revision. The developer provides service to users, along with a partnership in analysis, design, development, and implementation. The users contribute their expert application knowledge to develop requirements for a specialized information management system, the screen and database content, the system organization, and the specialized calculations.

Chapter 3

CONTRIBUTIONS TO THE SYSTEM BY THE DEVELOPER

Selection of a vendor or developer for an occupational health and safety software system is a decision to be made jointly by users and information services and is important to the outcome of the project. The vendor or developer must be experienced in software development and must understand some of the requirements of this specialized software.

There are methods and tools for software design and development that greatly enhance development of applications software and reduce development costs. Some of the software features the developer can use in occupational health and safety software increase interaction between users and the system, making the system appear user friendly. The product should be a menu-driven system with good security and database integrity. Table lookups, a data dictionary, linkages to facilitate movement through the system, well-designed screens, and programmed interfaces are features the developer will supply.

3.1 FINDING A DEVELOPER

Options. When you are hungry, you cook a meal or buy it, depending on the state of your pocketbook and refrigerator. Software acquisition is like having a meal. You buy it ready made, you make it, or you buy a package for modification to your own specifications. Depending on your pocketbook and other resources, the options are:

- Develop software in-house
- Purchase ready-made software
- Purchase software that can be modified

Selecting the best option requires research and analysis through a feasibility study and a requirements study, examining each option for cost-effectiveness and suitability.

Finding a Vendor. Trade journals for occupational health and safety are the best source for locating vendors. There are paid advertisements, and most journals have special sections on new products. Occupational health and safety trade shows are ideal for viewing software demonstrations. Consultants may have suggestions to offer, but usually it is the purchaser who finds the vendor.

Contact with the vendor will be followed by examination of marketing literature, which is useful only to the extent that it gives the reader clues as to whether the software is applicable. Not many vendors of integrated occupational health and safety software exist. This is a tough software market, and few companies devote their resources to it. There are more microcomputer special-purpose packages available, especially for MSDSs, hearing conservation programs, and other special uses.

The integrated system vendor will provide software demonstrations. If compatible machinery is not available at the purchaser's site, the vendor will gladly demonstrate the system in the home office. At any demonstration there should be representatives of all three occupational health and safety departments, in addition to information services.

Making Decisions. This is an expensive undertaking, so decisions should not be made quickly except under pressing circumstances. The delays frustrate vendors, but time is paramount for the purchaser who has to justify a large expenditure to top management, who has to live with the purchase far into the future, and who copes with annual (or longer) budget cycles.

An integrated occupational health and safety data management project will be a sizable company investment, whether purchased from a vendor or developed in-house. When no package satisfies all user requirements, cost limitations, and hardware compatibility needs, the choice may default to custom development, either in-house or on contract. Generally, it is more costly to use in-house resources, because additional personnel must be hired to dedicate to the project. Long development time and missed deadlines are inescapable risks.

The Importance of Special Tools. A key factor in the choice between custom development and vendor product will be the availability of screen and report generators. Ownership of these 4GL tools can save thousands of dollars in modification costs, both at the time of development (purchase) and in the future. The advantage will be significant savings in time and money if the tools are designed specifically for the needs of the occupational health and safety application.

Important Legal Considerations. When an expensive package is purchased, the buyer should check the provisions for obtaining source code with the software. Avoid the expenditure of thousands of dollars, only to have the vendor go out of business and no continuing customer support. Source code can be held in escrow against a vendor's business demise. If source code comes with software, there may be restrictions on its use, such as alteration of the code. Microcomputer shrink-wrap packages normally do not carry source code. This software may be copy protected, necessitating a run-time license for the number of occupational health and safety users. It is almost a certainty that special-application software will be copyrighted, and the vendor's limitations on copying and use must be respected.

An attorney (Harris 1985, 61) published a useful small volume called *The Legal Guide to Computer Software Protection,* from which the following is quoted:

> Subject to certain limitations, the copyright owner of any work has the exclusive rights to five important categories of actions regarding his or her creation. He or she may also sell or license these rights to another or make such other transfer of ownership as he or she deems appropriate. The copyright holder has exclusive rights to the following:
>
> 1. Public performance of the work
> 2. Public display of the work
> 3. Distribution of copies by sale, lease, rental, or lending
> 4. Reproduction of the work in copies
> 5. Preparation of derivative code

Despite the fact that users will depend on the company attorney to know about software legal matters, at the very least it is wise for users to be familiar with the legal restrictions on making copies. For those contemplating any use of source code, there should be a full understanding of their legal responsibilities. Contract negotiations should be under the supervision of the company attorney, with specific performance and acceptance conditions spelled out.

Option Limitations. Vendors sometimes lease their software, which is an option worth investigating. They may also offer on-line services (share systems). This may be temporarily useful during the transition to a new system in development. However, long-term share system use may be costly and not functional enough for most users.

No software options will be perfect. Trade-offs will be made to meet as many objectives as possible. The process of reaching a decision is always time-consuming, and numerous people from diverse company departments will have input into the project.

3.2 SYSTEMS EXPERTISE

Recognizing Expertise. Without computer expertise, how will users recognize it in a developer? There are no categorical answers to this question, but there are measures for evaluation. Some apply only to vendors, and others to both vendors and the in-house information services department.

Vendor Evaluation. The first question everyone asks of computer companies is "How long have you been in business?" This is supposed to determine reliability. It does not measure the excellence of the software; it only suggests the financial stability of the company.

In occupational health and safety, no integrated package has yet been produced that has become an industry standard. Financial stability is unrelated to expertise in applications software in this highly specialized area. A more reliable evaluation of the vendor's systems expertise in occupational health and safety is a demonstration of the system. Users quickly detect the degree of applicability of the system to their own work while viewing the content of screens and reports. However, they are generally too intimidated by computers to try hands-on testing.

Information services will determine the reliability of the software, explore the features it offers, and probe the methods used in development. Systems experts in information services may test the system and may request an opportunity for hands-on introduction to its operation. This provides some measure of reliability as they evaluate the system's responsiveness and dependability.

If the vendor has sold other systems, there will be users who can be phoned or visited. A visit to see the system in actual operation is preferred, but a telephone conversation can make or kill a sales claim about satisfied customers. A good question to ask is whether the vendor has successfully met deadlines and user specifications. If not, that should act as a warning.

Software Engineering. Whether the developer is a vendor or information services, the project manager should have software experience rather than hardware experience. The project involves software engineering, which differs from hardware engineering. As with any product that is engineered, the manufacture of software is based on specific knowledge of tools and methods and how they work together. In the broadest use of the term, software engineering covers analysis, design, and development of any software, but in Section 3.4 this term will be refined.

Referring back to the house-building analogy introduced in Section 1.1, the architect confers with buyers and makes blueprint drawings using his or her engineering background in load stresses and strains. The finished design is adhered to by the construction crew who actually build the house. Into every house go the same basic components such as footings, basement walls, joists, wall studs, and rafters. Once the basic house is built, special requirements are added to comply with the buyer's specifications for heating, plumbing, and electrical wiring. Finishing touches like color and decoration complete the buyer's specifications.

Software engineering is basically the same. No matter what the programming language, programmers begin with the system analyst's design, first coding basic algorithms to provide the software foundations. The first level of code lays out modules, programs, and menus. The second level of code is more specialized to apply to occupational health and safety, formatting the screen forms, making the system flow from the menus, and linking screens to an active database. The third level adds special enhancement features like help screens to the system.

The coding part of software engineering requires programmers who will follow "blueprints," rather than inventing a system they envision for the users. Users invent the system, the analyst designs it, and programmers construct it. Real expertise in programming is demonstrated by knowing how to read and follow the analyst's designs. Experienced programmers will know how to select the most reliable programming structures to fit the designated task.

The vendor or information services group with this kind of expertise will begin with a library of established subroutines and build from there. Screen generators begin with subroutines already in place, which is why a 4GL saves so much time and is cost-effective; it constructs the foundation code automatically and eliminates most of the debugging process at compile time. Once the basic system is written, unique features can be coded into the structure. This takes programming expertise, whether a screen generator is used or not. If a generator is used, the programmer must be especially adept at using the language of the 4GL tool.

Organizational Expertise. Whether a vendor or information services becomes the developer, applications development experience is important. The occupational health and safety system will have similarities to and differences from other applications, but it needs special features suitable to users' work procedures.

As emphasized earlier, a partnership must exist between developer and users. Bureaucratic software management will not work. How does the user identify the pitfalls of a bureaucratic developer? Ask questions to find out the following:

- How many projects are presently active?
- How many projects are pending, but not begun?
- How closely are these projects adhering to schedules and deadlines?
- How many projects are already late in delivery?

Explore the organization and work practices of the developer by obtaining information on the size of the organization and the number of levels of management. A large organization will have several or more levels. For each action or phase of system development, how many levels of review and approval will be required before proceeding to the next? How much paperwork is involved? If every plan, specification, schedule, and method has to be written, reviewed, and authorized by more than one level of management, users are going to experience a lot of time loss and frustration. This is why the size of the development group is important. The larger the group, the more management will be in force and the longer the project will take.

The occupational health and safety system needs checks and balances to ensure a degree of excellence, but it is too specialized and difficult to produce to get bogged down in politics and red tape. A small, manageable task force is needed for efficiency. Lots of enthusiastic managerial support can bolster development. Many small, informal meetings between users and systems personnel will take place. Too many people involved at any one time will blunt timely progress. Look for organizational simplicity.

Application Generators. While application generators make a significant contribution to the development of occupational health and safety software, a developer can produce a system without these 4GL tools if there are high-caliber analyst and programming services available. Application generators are cost-saving tools for the reasons shown in Figure 3.2.

At the same time, application generators (especially for large computers) can be expensive, and they may not offer all the elements useful in an occupational health and safety environment unless they are specially written for the application. In terms of costs, development of specialized application generators rapidly depletes capital, which is ultimately passed on to the 4GL purchaser.

Application generators offer an opportunity for improved systems production, and they are the newest technological tools. Nevertheless, successful software development is not dependent on tools and machinery. In the final analysis, it is interpersonal skills that spell success, which is why an excellent development team is a viable alternative to an application generator. The 4GL generator exploits generic subroutines by

COST-SAVING BENEFITS OF APPLICATION GENERATORS

Provide quick prototypes

Encourage user participation in design

Promote iterative design and review

Build around a database

Reduce programmer error

Fig. 3.2 The cost-saving advantages of application generators.

employing them extensively. The trade-off is that a percentage of this code creates memory overhead. Adroit programmers will know how to mesh subroutines and reduce the quantity of code at no expense to product quality.

3.3 APPLICATION PROTOTYPING

Prototyping. Prototyping is a software development methodology. Methodologies are sets of sequential rules and procedures used to solve a problem. In software engineering, there are methodologies for prototyping that follow approaches categorized as explorative, experimental, and evolutionary. The experimental approach is ideally suited to occupational health and safety. The essence of this is trial and error. If the user approves a prototype, it is used or further developed; if not, the prototype is discarded in favor of a new effort.

This works for the user interface, system architecture, and system functions. The user interface (e.g., menus and screen forms) is an example. Menus and screens are hallmarks of an interactive system; they are the means for communication between human and machine. Prototyping provides an opportunity for the user to participate directly in successful development through contributions to menu and screen design. An iterative dialogue between human and machine evolves as the developer learns more about user requirements and manipulates the prototype.

Prototyping is one of the newer methods of software engineering. It does not replace written specifications; it facilitates drawing up final specifications through an iterative process. Users continually change their requirements as they see the product evolve and as they contemplate the effect of a computer system on their work practices. What the user initially perceives as necessary is almost certain not to be what the user will eventually want, the shift being a developmental process. The user's potential to develop a system appropriate to his or her work can be cultivated through prototyping.

Models. A prototyped user interface is not a finished version of a system. It is a temporary working model, incomplete, and lacking functional capacity. Because it is experimental, with the potential for being trashed, it must be built quickly and inexpensively. If it is workable, it can be expanded into the completed system as functions are added. A workable prototype assumes, but does not guarantee, that the ultimate product will be a success.

As in many engineering fields, prototypes prophesy the success or failure of expensive structural models. Errors, large and small, are detected in their infancy before they mature into expensive corrections. The product that emerges from prototyping is a reasonable facsimile of the finished product, functional and reliable. In software applications, prototyping tentatively establishes favorable acceptance by the users. This is because the prototype has helped both user and developer organize the requirements of the system and define user specifications. Prototyping can be done on paper or computer, depending on the tools available. Screen and report painters support prototyping by setting up early models for user review in a matter of hours or days.

Levels of Review. The first level of review may be menus, screen forms, or both. On paper or on screen, the user will check them for prompts, labels, and data descriptions. Screen painters can direct the path of the cursor through the form, emulating work practices and thought processes. For instance:

- In vision testing, is tonometry done before color vision testing or afterward?
- On company records, are employee IDs, names, or social security numbers the primary identifiers?
- Are employees' nicknames shown on each record?
- How will screen titles read (e.g., "Air Sampling" or "Air Monitoring")?
- What is the correct sequence of programs and screens on the menus?

Iterative review keeps both users and developers busy, one providing input to the system and the other transforming input into a functional system. Every character on the screen is important to the development team, and users will be astounded at the extraordinary amount of detail they must evaluate.

The next level of prototyping will be functional. Screens will be attached to the database for entry, modification, and deletion functions. After that may come screen display retrievals for screen display and checking of execution time for efficiency. At the earlier levels, user decisions will predominate, but later levels are within the developer's province. By this time users will have experienced the system's operation and will have identified the features needing improvement.

Development Problems. While it is the developer's responsibility to produce a system to help users do their work, users need to understand the developer's problems. As an illustration of how a user can create problems during development, consider the corrective actions that follow an accident investigation. The user cannot guarantee a maximum possible number of recommended actions. Asking the developer to provide 15 lines (one action to a line) when only 2 or 3 are normally required is an unnecessary encumbrance of storage. The developer may propose a special programming solution for this kind of data, permitting any number of lines to be used without encumbering unused storage space. This creates another problem, however. If one line is deleted, how can all other lines in this element be left intact? The developer solves that problem with special deletion programming. Now the user has to remember when to employ the common delete function and when to employ the special delete function. Simplicity of operation has been traded off for a special requirement.

Frequently, when resolving problems, the solution slows retrieval and/or execution time. Will the user find slower execution time objectionable? Probably. Developers can occasionally increase execution time by additional programming effort, or the problem may be solved by optimizing the database (a method for organizing data retrieval paths for optimal speed), but the developer's power to resolve problems is not unlimited. For this reason, users need to remember a few simple truisms before requiring developers to change the system:

- More is not always better.
- Faster is not always better.
- Better can be worse.

Speed should not be the primary reason for implementing a computer system. More realistic reasons are improved recordkeeping and easier data retrieval. User expectations may need adjustment after weighing the trade-offs proposed by the developer. Users occasionally complain that they can write records faster manually than they can enter them into the computer. This may be true, but can the data be retrieved from paper files as rapidly as from the computer? Probably not. This trade-off is speed of retrieval for speed of entry.

Beyond Prototyping. Some parts of development are not possible in prototyping. For instance, batch processing requires special insertion into a system developed by a screen painter. This is not interactive. For instance, all the data from medical laboratory analysis (e.g., blood chemistries) and all the data from the analytical chemistry laboratory (e.g., air sample analyses) may be collectively entered into the database each night. Another example is running off employee notices of hearing examination results in a hearing conservation program. Still another example is locating all employees who should be scheduled for training courses of a specified type, for medical examinations, or for noise dosimetry surveys. This requires special programming, which will not take place during prototyping.

Prototyping also does not include coding for special procedures for security, periodic backups, or audit trails. The developer incorporates these features later in system development.

Disadvantages of Prototyping. Prototyping is a powerful development tool, but it has certain disadvantages. Users see something on the screen that appears to promise solutions to all their problems. At the screen review level, however, the system is not fully functional. When the system is in full use, they may discover unanticipated limitations. Users tend to become restive once they have tasted the fruits of development, which can be a disadvantage of prototyping.

3.4 MAKING A SYSTEM DESIGN

Good System Design. The importance of user participation during system development has been stressed. In systems analysis, users are helped to identify their information problems and needs. In prototyping, users examine alternative ways to perform tasks. At the same time, analysis and prototyping help the analyst to understand users' attitudes and work practices. Figure 3.4a shows some of the advantages to be gained by joint participation in system design.

ADVANTAGES OF SYSTEM DESIGN

Defines project limits

Controls project time

Controls project cost

Provides blueprints for programmers

Ensures product will meet user requirements

Coordinates efforts of project personnel

Establishes early project documentation

Provides documentation for later maintenance and repair

Fig. 3.4a The advantages of writing a system design for the occupational health and safety computer system.

With so much at stake, the necessity of good system design is obvious. A prevailing problem with system design, however, is the time it takes for design and execution of the numerous models that blueprint a system, especially a large system. The tedium and expense of manually drawing designs were previously mentioned. When the analyst has sketched the elements of design, they must be drafted. Either the well-paid analyst executes these final diagrams, or an assistant is hired for the task. The tools are inexpensive: a template, a fine-pointed writing instrument, a straight edge, and a lot of paper, coupled with many man-hours. Alternatively, a graphics package may be found to do the job when the initial input is time-consuming, but corrections can be made without redrawing over and over again.

Error-Free Designs. There is an absolute necessity for error-free designs. Mistakes easily find their way into the drawings. The reason is that broad (high-level) design of the system is "decomposed" into progressively lower levels of greater detail until programming can begin. When the

ADVANTAGES OF CASE TOOLS

- Reduction of development costs for drafting and redrafting
- Elimination of model errors
- Fast redrawing of designs to correct errors
- Cost-effective for large numbers of models
- Easy production of data models and data flow diagrams
- Foundation for system documentation
- Standardizes design for later repair and maintenance

Fig. 3.4b The advantages of using CASE tools for system design.

model is decomposed, sections are lifted from the larger model and each section is augmented for clarification. Each time decomposition takes place, the drafter risks dropping an element, misnaming it, or misdirecting the connecting lines. Additional time is needed for error checking and rechecking of details, which is described as quality assurance.

Software Engineering. Software design is an engineering process. Once the requirements for the project are agreed upon, the engineer drafts "blueprints" for production. There are two types of engineering related to software: software engineering and information engineering.

Real-Time Systems. Every plant today probably has computerized processing controls. Somewhere there is a gadget that measures equipment speed, process temperature, sheet thickness, or some other processing feature. The device sends its measurements to the computer, the computer compares the data with acceptable range limits, and when the data exceed those ranges, the computer sends a signal to trip an equipment switch or an alarm device. The design and programming for process control are termed *software engineering*. It is event driven and is called a *real-time* system.

Information Engineering. Information engineering, on the other hand, is data driven. This is strategic designing that begins with a single model illustrating the project in its corporate setting and its relationships within the company as a whole. Two-dimensional models illustrate the context of the project. Decomposition of this model separately describes each function of the system, with lower levels showing the data elements important to each of those functions and the manner in which they flow through the system. Elements are input, processed, and output as different data elements. They flow from their point of origin in the organization to their point of use. Information engineering is not affected by events external to the computer; it is the events internal to the computer that characterize information engineering used in system design. Experience in this type of design is derived from database management and business systems, not real time systems.

Methodologies. There are a variety of methods for doing *top-down* structured system design. Some are based on the functions to which the data are related; others, on the data elements themselves. These methodologies are object oriented, which means that the design techniques are based on units that can be shown as objects being acted upon in the computer system. The objects are depicted in the models as rectangles, ellipses, or round-cornered rectangles, as mandated by the rules laid down by the methodology inventor. Lines or arrows connect the elements to one another, showing the directions of input and output. An element label names the object or describes its function.

Methodologies are engineering disciplines with specific meanings to prevent designers from communicating in a personal manner. Even the number of levels of decomposition is declared by the rules, some methodologies prescribing three to five levels and others five to nine levels. An integrated occupational health and safety system will consist of several or more modules, each module will decompose into several or more functions, and each function will have many data elements. Hence, there will be many models in the hierarchy from the top diagram down to the programmming level.

CASE Tools. CASE tools offer newer techniques in system design and save a lot of time and money. The advantages of CASE tools are listed in Figure 3.4b. The trade-off is that CASE tools may themselves be expensive. Nevertheless, these tools save money in the end and provide quality assurance for the models.

CASE tools vary according to their vendor. Some provide systems analysis, some produce system designs, and some include features for both analysis and design. Nearly all are able to do graphics and to produce charts and diagrams. They check for errors of exclusion by tracing data elements through the design hierarchy, as well as checking for syntax and consistency errors. Hierarchical outlines from top to bottom catalog the locations of data elements, and the system automatically checks charts and diagrams against these locations. Some tools generate code from these designs, but CASE tools of this type cross over the border and become 4GL tools. 4GL generators are usually attached to database managers, which is not true of CASE tools. The difference between them is that 4GL tools generate code, while CASE tools generate graphic design. When CASE tools cross this boundary, they are joined to screen and report painters, generating formats but rarely code. The trend is toward automation of software development from analysis to implementation, including coding. Even expert systems (5GL) are now finding a place with CASE tools where the shells form a framework for analysis.

Integrated CASE Tools. The more sophisticated CASE tools are now packaged as toolkits that integrate a variety of tools. They can assist in prototyping by formatting menus, screens, and reports, and the designer can draw and manipulate the diagrams, storing them for future reference. CASE makes a distinction between storing data in a repository vs. a data dictionary. The repository includes information about data elements such as input origin, method of processing, and output destination. There may also be computations made during processing, security controls, and access privileges. Some people refer to the repository as the *encyclopedia,* or the *central dictionary*. In data dictionaries, attributes of data elements are described, such as the name of the element, what it covers, the type of element, and its size in bytes. (Section 3.13 will discuss data dictionaries in detail.)

The advantages of CASE tools are:

- Reduction of draft and redraft costs in development
- Elimination of model errors
- Automatic redrawing of designs to correct errors
- Cost-effectiveness for large numbers of models
- Production of data models and data flow diagrams
- Foundations for system documentation
- Design standardization for maintenance and repair

Labor-intensive design can be controlled by the developer who uses CASE tools, thereby reducing the cost of the project.

3.5 THE NECESSITY FOR A MENU-DRIVEN SYSTEM

A Menu-Driven System. Users need to navigate quickly through the system. The menus that serve this purpose are like file cabinet drawers that separate personnel data from payroll data and current employees from terminated employees. Terminated employees are further divided into retired employees and those who left to work elsewhere. In the computer, menus and submenus replace labels on file drawers and card separators within the drawers. Select an item from the menu, and the system carries the user directly to that file. In a menu-driven system, database files are linked to the menu.

One of the first steps in system design is to divide the system into manageable functions. Some data management systems have been divided into functions like data entry, data modification, and data deletion. This forces the user to return to the menu when shifting between functions. Designs like this are archaic and reveal a developer's limited programming skills.

Menu functions are best linked to user work practices and database partitions for occupational health and safety. Chapter 1 mentioned partitioning of the database for industrial hygiene, the clinic, and safety management, giving each department control over its own section of the database. The employee database is another separate division and would be available to all three departments. MSDSs could be another separate division made available to everyone. With divisions like these, the top menu might show:

Employee Information
Clinic Management
Industrial Hygiene Management
Safety Management
Material Safety Data Sheets

Menu Levels. Items on the top menu would, in turn, lead to subfunction levels. In occupational health and safety, subdivision of work functions readily subdivides the database as well. In Section 4.7 examples of further breakdown of the top menu will be given. In this section, only the principles of organizing menus will be discussed. During analysis, the developer guides the user through the process of breaking down functions in order to construct menus that follow work routines.

Modular Design. A modular system design will make a large, integrated system more efficient to program. Modularizing also makes it easy for programmers to get into programs for maintenance and repair at a later date. The menu items listed above illustrate modules. The next step is to divide these modules into programs. Each program will represent the work routines within a department. For instance, industrial hygiene functions might be surveying, making exposure calculations, and keeping track of industrial hygiene equipment. Programs can be further divided into specific tasks. Surveys may be made for air and noise. In each case, there are data specific to the task, making screen forms at the next level. Thus, the levels would be modules, programs, and screens.

Within a screen, all functions for data entry, modification, and deletion should be available but not linked to the menu. Users will furnish titles of programs and screens to the developer for menu construction. This creates instant user recognition upon implementation of the system.

The system should be driven by menus written to follow the user's work routines in order to structure the system for the user. Change disrupts, and the system should conform to existing work practices in order to complement those practices. An occupational health and safety menu system should be flexible and modifiable because, as personnel changes take place, work practices will also change.

Menu Item Selection. The means for selection of menu items can also be a developer's contribution to a good system. This is another feature that measures a developer's expertise and indicates whether the developer keeps abreast of advances in computer technology. In the past, for instance, selections were made by entering a number (1, 2, 3) or a letter (A, B, C) next to an item on the menu. The drawback is that this method requires a user to locate specific keys on the keyboard, and if an error is made, the system is sent to a wrong file. Technology overcame this "hunt and peck" method with the advent of touch screens, icons, and the hand-held mouse. Touch screen selection turned out to be a passing fad, but icons and mouse methods survived the rigorous test of user friendliness. These were innovative developments born with Apple Computer's LISA. LISA perished, but icon and mouse methods lived on.

Pull-Down Windows. Another innovative development was *windows*. The user does not have to leave a screen to display other data or programs because, at the touch of a key, a *pull-down* window appears on the same screen like a window shade being unrolled. This is a programming technique that permits many user-friendly features—pull-down menus, for example.

Pull-Down Menus. Menus placed in windows are user friendly. For mouse-driven systems, a mouse clock at the right place makes the selection. For all other systems, selections can be highlighted as the cursor is moved down the list, with the user pressing the Return Key at the preferred line. A menu-driven system with simple selection methods like these:

- Makes selection easy for the user
- Reduces key strike errors
- Minimizes user eye strain

Windows should vanish automatically upon menu selection, putting the user into immediate contact with the module, program, or screen to be used. A series of pull-down menus activated in quick succession by the Return Key will speed the user through modules, programs, and screens to a correct file destination with a minimum of effort.

3.6 AN INTERACTIVE SYSTEM

Intercommunication. *Interactive* is a description of the communication between human and computer. The relationship has subtle qualities that are overlooked by users. Like the importance of gravity to our earthly existence, we tend to take for granted the phenomenon of human-computer interaction. It is the subtle features of both humans and computers that a programmer takes advantage of to enhance communication.

An automobile and its driver are a good analogy. The dashboard communicates a variety of messages to the driver. The most common one is the rate of speed; less often used but frequently read are the gas gauge and the odometer; in an impending emergency caused by low oil or an overheated engine, a red warning light will flash; other assorted messages are also conveyed by dashboard signals. All of these devices send messages to the driver about the state of the vehicle. In response, the driver directs the vehicle to accelerate or decelerate, stop for refueling or repair, or whatever it takes to control the driving operation. The video display screen is like the automobile dashboard because it keeps the user informed. The human uses the video display screen as a dashboard and the keyboard as the steering wheel.

Interactive Language. As discussed in Chapter 1, the computer responds to off-on switches in its electronic circuitry. The human circumvents actual switch manipulation by instructing the computer to set the switches through machine language. Machine language, in turn, is activated by high-level languages to direct both processing and machine operation by translating human commands into machine-readable commands. Both the keyboard and the video screen also have signaling devices linked to the computer machinery that provide a pathway from keyboard to computer and from computer to video screen. An interactive system provides a continuous communication between human and computer.

Signal Devices. In addition to the language pathways, computer devices —the keyboard and the video screen—are employed to receive and transmit messages. This is the interface most apparent to users. Also ready to receive are human eyes to accept input, send input to the brain, and neural responses to output through the fingers. Consequently, it is the combination of human and computer interaction that is the real system. By itself, the machine is nonfunctional.

Certain capabilities of the machine have been exploited by programmers to attract human attention to the machine. Sounds can be used to stimulate human hearing, while eyecatching features stimulate sight. Examples are:

- Bell (buzzer) that is an amplification of the computer's electronic signals. It usually beeps as a warning signal.
- Increased or decreased light intensity on the video screen. Increasing the light behind screen characters is called *backlighting* or *highlighting*. It focuses the eye on an area of the screen. Increasing the light intensity of the characters instead of the background is called *bolding*.
- The cursor that tells the human where the machine is ready to receive input at any given time is the signal device most commonly known to users. It is the most important user signal and the one taken most for granted. Cursors can be an underscore or a block; their position is noted by enhanced light intensity and/or blinking.
- Each key on the keyboard sends a unique message to the machine. Some of the keys are alphabetic, some are numeric, and some serve special purposes like the backspace to delete a character, the Return Key to signal the end of a line, or the space bar to signal the end of a word. There are function keys to perform special functions (macros), some of which can be programmed by the user for personal use.

Feedback. Although users overlook the process, the computer has been programmed to feed back instructions from the human for verification.

Every character entered at the keyboard is sent to the video screen for verification. How else would the human know when an error was made?

Humans become aware of feedback when the computer balks at an instruction because it is unable to perform as instructed. This leads to human frustration and breeds distrust of the computer. Everyone likes to blame the computer for an incorrect bill, all the while knowing the real culprit to be the human who programmed the machine. Machines have exacting requirements for their instructions. Communicating effectively and accurately with the computer through programming languages is an art and a skill. Creative programmers exploit machine capabilities. Some create innovative techniques for communicating with humans, such as pull-down windows and pull-down menus (Section 3.5).

Status Lines. A good application program will carry status lines on the screen at all times. This provides constant feedback to users about computer conditions. In occupational safety and health systems, status lines can describe which module, program, and screen the user is on, which variable is being entered, what kind of variable it is, and what it has to offer. For instance, are there two or more coded values that can be used on this variable? If so, the user can pull down a window to show all possible choices and make a selection. If there are a number of lines of text available, it will be helpful to know how many have been used and how many are still open. Other status lines can tell the user if this is a new record for data input or an old record that will only accept modification or deletion.

The variety of information contributed by the status line should be determined in development by both user and developer. Users may also need information otherwise hidden in the computer, such as the date and time of the record input, change, or deletion. There may be stored data relating to the screen that the user would not know were available without a message. If the computer is unable to respond to user directions, error messages can be shown on the status line. Status lines are valuable dashboard signals once the user gets in the habit of using them. If the status line contains a lot of information, the programmer may save space with coded or abbreviated messages. This can slow the learning process, so the status line messages need to be evaluated for genuine need.

Programming an Interactive System. It is in programming the interactive system that another measure of development expertise can be taken. Throughout this part of the project, programmers have to sense the pitfalls that can trap users. Warning devices should be employed to send clear messages about the nature of the problem to users. The most typical response is an English statement called an *error message*. A glossary of error messages usually appears in the user manual. Good manual documentation will describe a remedy for the error condition in order to allay user frustration.

The experienced programmer uses error messages to promote interactive resolution of human-computer problems. The programmer sets up a friendly situation for the user that prevents the system from stalling or crashing, conditions from which there can be no recourse except to shut the system down. Error messages attempt to direct users into a path that will maintain an operating condition and permit the work to go forward.

User Friendliness. The computer cannot think, but it can test against conditions, a form of logic in which a condition either exists or does not exist. When directed by sequential commands, the computer can algorithmically process instructions in its CPU. The computer can send bytes (computer word lengths) of data to magnetic media for storage. The machine has very limited capabilities, but when a programmer organizes simple instructions in the proper sequence, a great deal can be accomplished. It is the programmer's responsibility to remain sensitive to problems that the user may encounter in order to keep the system operating evenly. If the truth were to be recognized, a computer system is user friendly only when a user-friendly programmer has made it so.

Making systems user friendly takes programming time, thereby increasing the cost. Wisdom and experience have to be employed by the programmer in promoting user friendliness. The system can be overloaded with user-friendly features, increasing the response time needed for the machine to test the status of conditions before proceeding to the processing function. By itself, processing consumes time, and any additional time must be weighed for its value (cost vs. contribution). Too much user friendliness can irritate the knowledgeable user. Hence, the user may be expected to forfeit some user-friendly features to gain an optimally functioning interactive system. A wise developer helps the users to understand the necessity for keeping user friendliness within appropriate limits.

3.7 SYSTEM SECURITY

Entire books are written on this vitally important area. Because so many kinds of computer system security can be provided, the intent here will be to suggest the scope of security ideas that the user and developer can discuss during design and development. Security is designed during development; it cannot be implemented as an afterthought.

Physical Security and System Security. For the occupational health and safety application system, design should include both external (what happens to the system) and internal (what happens within the system) access controls. System security protects the organization against intentional misuse of the hardware, software, and database. C. J. Date (Date, 1986, 487) says that security ensures "that users are allowed to do the things they are trying to do," while integrity ensures "that the things they are trying to do are correct."

There are two aspects to computer security:

- *Physical security*. This is the prevention of physical destruction of the equipment (hardware and software) or the database. Physical security protects against fire, burglary, power failure, equipment failure, and disasters such as flood, earthquake, and hurricane.
- *System security*. This is the prevention of incorrect or lost data because of system failure or personnel errors. Included in system security is database security. Any alteration of the true data, intentional or unintentional is a direct security threat to the company.

Design for Security. The prevention of fraudulent computer use begins with controls designed into the system. The system designer must be sensitive to the value of the occupational health and safety database to the company economically, legally, socially, and ethically, incorporating the necessary security to protect these interests. Any physical damage to hardware or software may result in loss of vital data and the unauthorized disclosure of proprietary information. Whether the cause is accidental or deliberate, direct or indirect, large or small, the company is threatened to some degree. In the worst case, computer fraud can terrify company management.

Security begins with the selection of the computer, machine architecture, and operating system, providing the foundation for making a computer secure. The systems analyst will provide a report on how much security the equipment offers and any additional security requirements needed. Predicting security is a juggling act because absolute security, if possible, is prohibitively expensive. Consequently, the analyst will survey the risks for different kinds of security breaches, prioritize them, and assess their dollar impact. Analyzing the cost of potential losses compared to the cost of protective measures and procedures is called a *security risk assessment*. Risk assessments are difficult because dollar values for intangible assets like data are a matter of opinion. Occupational health and safety data is sensitive material, some of it confidential (e.g., medical records), and its loss or distortion can lead to expensive litigation or costly government citation. Security risks vary in every computer system environment, providing no easy answers, although users typically want maximum protection. It is up to the company to set the limits for security. While complete security may be costly, reasonable security is cheap compared to the costs of litigation. Therefore, an optimal balance must be found between security costs and company requirements.

Personnel and Security. Anyone using the occupational health and safety system becomes a steward of the company's well-being and assumes responsibility for the security of its data, abiding by all company policies and procedures that protect system security. Some personnel threats are shown in Figure 3.7a.

Privacy and Confidentiality. Security protects both the company and its employees and ensures privacy. Privacy is a constitutional right given to individuals and organizations to determine what information they are willing to share with others. The Privacy Act of 1974 established rules for federal agencies to follow regarding information about individuals, and many state governments followed with their own laws. Private organizations were not included in these laws, but courts have tended to expect companies to adopt policies in regard to information about their employees. The legal position is that individuals should be able to access

PROBLEMS IN MAINTAINING SECURITY

Programmer errors and omissions in
  analysis, logic, design, and coding
Carelessness in system operation, such
  as mishandling tapes and disks, and
  poor attention to system operation
Disgruntled or dishonest personnel
Lack of respect for the privacy and
  confidentiality of the database

Fig. 3.7a Problems in maintaining computer system security.

information about themselves, and be able to correct inaccuracies and misrepresentations.

Disclosure of information relating to personnel is limited to internal use. Privacy entitles the individual to either extend or withold information. For instance, under contract, unions representing employees can obtain information. The company clinic can release medical information to personal physicians and hospitals if the employee gives written consent. The salient point about privacy is that the individual makes the choice, not someone else. This is why groups want your signed approval before releasing your name on their mailing list.

Confidentiality differs from privacy in that it is a status accorded information of a sensitive nature. It protects privacy and leads to the necessity for database security. This limits the persons who can have access to occupational health and safety data and specifies which data each person is entitled to access by virtue of his or her job responsibility.

Types of Control. A wide variety of controls may be considered, but only a few may be needed. The motivation for system security or any of its parts are given in Figure 3.7b. The questions that arise will center on What, Who, and Which (in that order). Or on Things, Groups, and People.

The place to begin is at the company level and existing policies. Most

REASONS FOR SYSTEM SECURITY

Legal, ethical, and social implications

Corporate policies

Control over operations and production

Protection from theft and/or vandalism

Misuse of database

Protection during man-made disasters

Protection during natural disasters

Fig. 3.7b Reasons for establishing computer system security.

companies have written policies relating to security. More specific policies for computer security should be derived from these sources. The company will want to protect not only against property loss or damage, but also against the possibility of personnel actions. Any system is only as secure as the honesty and good intentions of its users. Policies should be included in employee training, and the consequences of accidental or intentional security breach should be clearly stated. Users are the first line of defense, and most are willing guardians of security. However, disgruntled employees have been known to sabotage systems, and unfortunately, the potential for this always exists.

At the minimum, security policies and procedures should cover protection of the database against unauthorized modification by:

- Identifying users
- Verifying data correctness (see Section 3.8)
- Controlling disclosure of database contents

Two means for control are security technology and informed users. Technology must be evaluated for its applicability, cost, and ease of use. Security devices must not impede use of the system to the degree that they slow down or frustrate users. The occupational health and safety system is intended to support surveillance in the workplace, not impede it.

In the area of disaster control, there should be plans for ensuring continuity of the database in the event of disruption, just as plant employees are trained to shut down and lockout in emergencies, to restore normalcy, and to start up again. A final caution about controls is not to overlook peripheral loss of data by other methods of tampering. Electronic mail can lead to insertion of viruses, as can numerous external routes through modems and outside access to the system. Printers, plotters, and any other computer peripherals should be controlled, along with waste baskets holding confidential data output, key cabinets, file cabinets, and documentation.

Security Methods. There are ideal commercial products and internal procedures, depending upon the situation. Commercial products protect against intruders, interception of electronic signals, and machine abuse. Commercial warning systems protect against smoke, fire, water, breaking and entering, and power failure. User-identification products are available to scan signatures, compare fingerprints, and/or recognize plastic IDs.

User identification is the most universally employed method and takes many forms; user authorization and passwords are the most common ones. Several levels of authorization may be required to enter the system and obtain access according to the tasks the user performs. Commercial products may be combined with all of these.

An up-to-date inventory of all hardware and software is a good beginning for system control. This includes keeping track of copies made of software to prevent fraudulent use of the copyright (see Section 3.1). The inventory should include the location, make, model, serial number, date of purchase, vendor, cost, and warranty dates for both hardware and software. In addition, information on who uses the equipment and what kind of security exists is advisable. Updating the list may be a nuisance when microcomputers and diskettes are easily moved from desk to desk and room to room.

Inventorying and access privileges are procedures for which the cost is limited only by the time they take to complete. No special purchases need be made. Procedures can be as simple as a sign-out list next to the door as people enter or leave the computer room. The point to be made is that security need not require special equipment. A small investment in everyone's time may be sufficient for effective security.

Hardware and Software Methods. While people are checked by identity, computers are checked by configurations, special electronic signals, software protocols, and special test programs. Computer communications have protocol checks to protect against transmission errors.

Encryption is a frequently used method and an expensive form of security. If files are encrypted, they can be declassified, presumably because cryptology sanitizes the contents. Encryption does deter most inquisitive people, and breaking codes may require several lifetimes of trial-and-error to hit the winning combination. Encryption also requires decryption, meaning that both encryption and decryption keys must be kept from the wrong hands. Encryption should be employed where intense security is needed, such as in public telephone lines. In a computer system, encryption will probably increase the response time, even though the cause will not be apparent to the user.

Security methods are best if they are easy to install and transparent to the user. At the least, they should not interfere with users on the system. Too many locks and keys irritate users. It is better to use subtle methods of protection such as not displaying menus on the screen until after log-on has been completed by the authorized user. Make frequent backups of the database and store them in a safe place, dated and described. Remove all diskettes and disk packs when computers are not in use and keep them under lock and key. The location of hardware is a security concern. Offices that can be locked at closing time deter theft and tinkering, but this may not be possible if the maintenance crew comes in to clean after everyone goes home.

Microcomputers are an invitation to theft. They can be bolted to desks with special equipment. Key-operated locks can be installed over the system switch to keep out unauthorized users. (But don't leave the key under the mat!) Locks can be installed on disk drives, and hard disks can be protected. (See "Microcomputers" below.) Floppy diskettes can be encrypted or kept under lock and key. If microcomputers are linked to larger computers or networks, they should receive special security attention. Any downloading from main computers to network workstations or to microcomputers decreases mainframe security, and the system becomes vulnerable to tampering and theft. The locations of microcomputers linked to larger systems should be under tight security.

Personnel Identification. Personnel identification techniques measure a variety of biological and personal identifiers like fingerprints, hand shapes, voice patterns, and eye retina patterns. One story of doubtful authenticity is about the company that measured phrenology (characteristic bumps on the head) by a helmet lined with sensors. A chin strap snapped shut if the user did not pass the test. When it came time to test this security marvel, a company with about two dozen data entry operators was chosen for the trial. At the moment of measurement, a power failure caused all tests to fail. An engineer summoned to the emergency found all two dozen people chained to their computers with helmets on and chin straps locked. The moral of this story is to find user-friendly techniques.

Another personnel identification method is to measure the shape and timing of signatures in comparison to a signed template. The problem is that measurements are not always identical for any one person. The movements of a cold hand may be jerky; a domestic quarrel with a spouse before leaving home may result in tight script; a two-martini lunch may lead to a scrawl. To overcome these variations, statistical methods attempt to decide whether the measurements are within acceptable limits.

Fingerprinting is done by placing an index finger in a cupped pad and entering an ID. The machine scans the print for the right features to match against the print on record. Of all the biometric methods, this one seems to be less threatening than most for users. Biometric access control machines can be costly, but commercial products are especially good for secure areas. In general, equipment like this is used only in tight security areas like military compounds, intelligence situations, and national security.

Passwords are the most common form of nonbiometric identification.

When entered into the computer, they must not be echoed (displayed) to the screen. They need to be changed often, and the security administrator should be the keeper of all passwords. Indeed, the security administrator may need to notify users when to change passwords. These words are selected because they are easy for the user to remember, like a favorite hobby or something familiar. Passwords like this are great fun to guess if you know anything about the user's personal life. Time and time again, passwords are broken because users do not select passwords ensuring maximal protection. The safest method is to have the database or security administrator assign passwords to users.

Types of Access Control. Access is controlled according to the user's identity, the user being an individual, a group (e.g., department), or a particular job assignment (e.g., a nurse, but not the nurse's aide). Once access to the program is obtained, there must be control over what users can do with the database. There are two ways in which these controls can operate: by object (object-oriented) or by function (action-oriented). Examples of the objects over which control operates are listed in Figure 3.7c.

The more controls, the finer the access distinction, and the more cumbersome use of the system becomes, which increases operating overhead. Therefore, trade-offs must be weighed between the optimal amount of security necessary and processing efficiency.

Action-oriented control is based on database functions to be used such as:

- Read only
- Read and write
- Modify
- Write only
- Delete

Reading and/or writing are common to most database users. When coupled with objects (i.e., modules and files), they constitute the majority of access control. Most write privileges permit changing or correcting data. In some instances, it may be desirable to have additional controls; modify may illustrate this. In the industrial hygiene module, air monitoring results may occasionally need adjustment in the professional judgment of the hygienist. Only the hygienist would be granted access to change (modify) these special fields. Deleting data from a record must be done with maximum care. In some cases, deletions may be restricted to specified personnel, but in most cases, verification is sufficient to protect

```
OBJECTS FOR ACCESS CONTROL

    An application program
    A database management system
    A database
    Files
    Fields (variables)
    Screen forms
    Modules
    Programs
```

Fig. 3.7c Suggested objects in establishing access control.

the database. When deletion is requested, the programmer puts in the message "Are you sure? (Y/N)." This causes the user to reconsider the action before making it final.

Audit Trails. Audit trails are an essential software feature of mainframes and minicomputers that track system and database use. They are especially useful in determining what went wrong in a transaction process. They provide major evidence of attempts to violate system security by recording the sequence of transactions. An audit trail can include information like the following:

- Log-on/log-off time and user identity
- File access, user identity, and file encryption
- Date and time of the transaction
- Administrative actions taken by the database manager or the security administrator

Audit trails are available only on some microcomputers. They are not typical because the amount of memory and storage space they require prohibit their use without a very large capacity hard disk.

Backups. The one security feature that is common to all computers and users is making backups. On larger machines, this is a regular task of operating personnel; on microcomputers, users are trained to make at least daily backups of their work. Backup tapes and disks should be stored in a secure place. Many companies keep theirs in bank safe deposit vaults. Each week they put new tapes in and remove the old ones.

The cheapest and most important security is that which is practiced on a daily basis, like personal hygiene. A backup is simply making a copy of your work frequently so that you can avoid losing any data. Just as some people brush their teeth after every meal, and others only in the morning and evening, computer users develop the habit of backing up their work either several times a day or once a day.

Most computers are reliable machines, but glitches do occur in their operation and files can be damaged or lost. Tapes, diskettes, and portable media can be misplaced or stolen. Since data are an asset, just like money or a wrist watch, the company will find ways to secure them. The first line of defense is the backup copy. The second line of defense is a secure storage place for both original software and backups. By cultivating the minimum habit of daily backups, the most that can be lost is a single day's work.

All software originals should be copied and the originals stored in a secure place like a bank vault. A copy is kept under lock and key for use, if necessary. A second copy can be used if it is necessary to install the software frequently.

Working data can also be protected by the two-copy method. Make two backup copies of the data to start, and label them A and B. Put A in a secure place and keep B for backups for that week. At the end of the week, put B in the secure place, remove A, and continue to use A for backups during the following week. Replace the secured B with A, and continue alternating weeks with backup copies.

Security Administration. The authority for access control of the occupational health and safety system should be delegated to one person, such as a database manager or database administrator. In a very large organization, there must be security oversight for the application hardware and software, as well as for peripheral devices like printers and modems.

During design, the systems analyst will suggest whether a security administrator will be needed. The security administrator grants and revokes access privileges, provides for periodic changes in passwords, and establishes and reviews physical controls and verification procedures. When the administrator grants access privileges, the responsibility is delegated to positions, not individuals. When individuals move vertically or horizontally in the company, there is no automatic transfer of access privileges that go with them; privileges remain with the job.

Starting with the analyst and programmers during development, the administrator provides information about the physical locations of computer, terminals, and peripheral equipment and assists in determining the locations to be made secure, especially where microcomputers are distributed throughout the company. Estimates of losses of computer equipment are staggering, according to news reports, with annual totals running into millions of dollars. Much of this loss is preventable through proper security. Even small decisions like how many times to attempt log-on before a user is turned away are part of the development process. Creating security is filled with decisions, both small and large.

Each user of the system needs to perform certain tasks within the system, ranging from data entry to report preparation to system queries. The security administrator has to understand each job function in order to make certain that users have sufficient access to the system to carry out their job. In excess of that, security dictates that no further right to use the system is necessary.

There are three levels of computer rights (Moffett and Sloman 1988), as given in Figure 3.7d.

COMPUTER RIGHTS

Company - right to own the system and its data

Security administrator - granted right by the owners to oversee the system for the company

Users - granted right by the security administrator to enter the system to fulfill work assignments and to view all or designated parts of the database

Fig. 3.7d Three levels of computer rights.

Mainframes and Networks. Large-computer and network security covers a variety of users and operations personnel. Policies and procedures must be followed by everyone after initial training.

Networking is a real problem, and users must be identified carefully. One typical procedure for user verification is callback. The user enters his or her ID and password and signs off; the network calls back, the user responds, and system access is completed. Network workstations need good physical security, beginning with protection at the communications port of each station.

Microcomputers. The most difficult security controls are for microcomputers because of their vulnerability to theft and vandalism. Unlike larger computers that require several or more personnel to run the system, microcomputers are one-person operations. That person has to be a reliable employee. The design of microcomputers is called *open architecture,* which means that the design is open to any access, authorized or unauthorized, with no accountability for user privileges and no password use before authorized access takes place. However, OS/2 systems now provide for keyboard passwording at the time that the software is accessed.

If a microcomputer is part of a LAN or a distributed system, it can be a source of database misuse by anyone able to download from the mainframe, minicomputer, or network.

Disk drives and hard drives are vulnerable to tampering. Hard disks now store a lot of data. There are two ways to protect them. For microcomputer users concerned about protection of disk drives, it may be worthwhile to consult a computer security person about the most effective control for the specific situation. Another way of protecting microcomputers is to use Bernoulli boxes and keep all software on Bernoulli cassettes. Remove the cassettes at night, lock them up, and reinsert them in the morning. Two-drive Bernoulli boxes are expensive, as are the cartridges, but they can replace hard drives very effectively and provide an extra measure of security.

Avoid unnecessary duplication of the database, but make backups and keep them in a safe place. Control is lost when too many copies are floating about. An unseen control on diskettes is that things are not totally deleted, which is one way to detect software tampering. Widely used utility programs in expert hands are able to glean hidden information that tracks unauthorized use of the software.

Power surge protectors on microcomputers will protect against power damage.

3.8 DATABASE INTEGRITY

Data Integrity. In Section 3.7, it was stated that system integrity ensures that the hardware and software function properly under the given security system. In this section, the emphasis will be on ensuring data integrity, namely, that the data entering the system will not differ from their original form and will be protected from loss or destruction. Data must be consistent with source documents, validated, and verified. Database integrity is implemented from within the system rather than externally.

Designing for Database Integrity. The designer's attention to the detail of data entry and processing will establish database integrity. The chief problem will be to avoid over design, which inevitably increases data entry time with small delays at each check. The quality and quantity of data checks must be compared with the importance of the data. Vital pieces of data may justify delayed entry, but in general, this is not necessary.

The security administrator will work with the developer to establish data consistency, validation, and verification, and the effort will be ongoing. As the system grows, codes have to be maintained, and policies and procedures for maintaining database integrity will require periodic review and updating.

Data Consistency. When data are entered from source documents, there is always an opportunity for error. Eyes misread and fingers slip, and it is common to destroy source documents once the data are entered, leaving no way to check a suspected error. The objectives in data entry are to prevent loss of data, to avoid making unintentional errors, and to verify correct transcription of data from source to computer. This is a tall order for mere humans!

One way to improve data integrity is to input data directly from the machines making the measurements, as during spirometry, audiometry, air and noise monitoring, environmental monitoring, and laboratory machine analysis. The day may come when these data are automatically sent to occupational health and safety computer systems, but at this time there are technological limitations.

Other methods are also being developed. Scanning documents is not a guaranteed process, but scanners are constantly improving. An area in which this will save data entry time is recording medical histories and updating them, provided that the data are entered in a form that can be read by a scanner. Until direct entry and scanning are perfected, however. transcription from source document to computer requires careful data verification.

An argument for data entry by nurses, technicians, and professionals is that they are familiar with the data and will instantly recognize some errors. There is a lot of truth in this, but these people are not paid to spend their hours entering quantities of data; their energies are needed for technical responsibilities.

Each person entering data will describe a fact or condition differently from the next person. To maintain data consistency, it is best to codify information throughout the database so that data mean the same whether they are in occupational medicine, industrial hygiene, or safety. For example, job codes must be uniformly used throughout the plant. Departments, locations, and processes must also be consistent throughout the database. Codification of data will be discussed in more detail in Section 4.10.

Data Validation. Data validation and data verification are capabilities that the developer can design into the programming of the system. *Validation* will determine that database integrity is maintained. There are a variety of validation procedures.

Each file has key words that make it possible to retrieve unique data efficiently. At a minimum, key words have to be entered to store a record. The programmer can require as many fields to be entered into a record as desired before the file is committed to storage. This is a completeness check on the record prior to storage.

When transactions are processed in a batch, the system should be programmed to validate the totals to make certain that records were not missed. The log file will tell how many records were processed, where they came from, and what became of them. All totals must tally in a validation check.

Standards can be set for determining the reasonableness of the data. The system can check for duplicate records and refuse to accept them. A sequence of processes can be checked against a standard to make sure that they were correctly done. When reports are run, checks can be made to account for the number of records available against the number processed.

Data Verification. To ensure database consistency, the developer can program checks into the software, a process called *data verification*. Before processing or before committing data to storage, the programmer sets up comparisons that "edit" for features like those shown in Figure 3.8.

```
DATA VERIFICATION FEATURES

Alphanumerics
Numerics
Blank fields
Length of field
Position of decimal point
Field format
Range
Existence of previous records
Consistency between similar fields
```

Fig. 3.8 Features used for data verification.

A blank field should be checked for several equally important reasons. The first reason is that it may be a key field. Lacking the key, a record is not storable and can be lost. The second reason is that a blank field is missing data. If the blank is unavoidable, the system should tag it as missing. When data are retrieved for a report or statistical evaluation, programming should direct the system not to include this entry in a tabulation, but to tabulate all missing data fields separately. A blank numeric field should not be counted as zero. Zero numeric data should be legitimate only when the zero is an actual value.

Field formats check entries like social security numbers that carry a full nine digits. The use of hyphens is an option that the user specifies during development. Hyphens and slashes are called *delimiters* and make a number easier to read. Verification determines that the digits are grouped correctly. The system should not accept an entry until the field is formatted correctly. Another normal use of delimiters is the separation of the month, day, and year by hyphens or slashes.

Numeric fields are formatted for real or integer numbers. Real number fields require uniform placement of the decimal point. Without this, report totals will be distorted. In transcribing the number from source document to computer, it is also helpful to be able to read the sequence of digits on either side of the point.

Data entry has its risks. The best way to improve data consistency is to recheck every entry against the source document, but the tedium of this operation and the pressure of getting a job done will overcome the best intentions. This is not a method of data verification that will be practiced on a daily basis.

Programmers make compromises to achieve optimal data integrity. If the system is tightly programmed for data verification, some user friendliness will be lost. Conversely, without verification methods, the data will be error prone. Thus, the extent of data verification to employ must be weighed. The burden of protecting the company's occupational health and safety data bank must be borne partly by humans because the machine cannot be relied upon for sole verification of the data.

Database Recovery. Mainframes and minicomputers use recovery methods to prevent loss of data during transaction processing. The amount of overhead they require precludes using the same methods in microcomputers. Recovery methods make use of transaction logs by recording every execution of a processing command that affects the database. These files are stored in compressed form to reduce overhead. If the system fails, restoration of the processing can begin up to the point of failure. Rollback procedures are the most common recovery methods used to prevent erroneous entries into the database.

There are four steps in rollback recovery:

- Rerun each completed transaction up to the point of failure.
- Abort any transaction in process at the point of failure.
- Remove any database alterations affected by the abortion of a transaction.
- Restart the processing of aborted transactions and complete them.

3.9 THE IMPORTANCE OF TABLE LOOKUPS

Table Lookups. In an integrated occupational health and safety system, the amount of data to be entered increases proportionately with the number of employees. For a company of only a 100 employees, input will be time-consuming if all the records for occupational medicine, industrial hygiene, and safety are counted. In addition to the time required and the increased data entered, the opportunity exists for data entry error. The best control over time consumption and entry error is table lookups. These are special files containing frequently used information that provide standardized data across the plant and corporation.

An easy way to understand the effectiveness of table lookups is by illustration, using the most common data in the database: employee name (last, first, and middle initial), employee ID number, and social security number. Entering all three items about the employee completely will require approximately 56 characters in every record about that employee, assuming 35 characters for the name, 10 characters for the ID number, and 11 for characters for the social security number. If 20 records are entered on each employee every year, that would total 1120 characters entered. For 100 employees, this would amount to 112,000 characters just to enter the most basic information. Disregarding time consumption, the opportunities for typographical error are myriad, resulting in a loss of database integrity.

By committing the employee's name, ID number, and social security number to one alphabetic listing based on the last name, record entry can be completed by the system rather than the user. Entry of the ID number sends the system to a lookup table which is sorted numerically, and the name and social security number are automatically returned to the user's screen. Once the employee information is accurately entered into the system, consistency is ensured thereafter through table lookups. If the ID number is in error, the system returns either the wrong name or an error message, giving an immediate signal to the user. This key table can be the name, ID number, or social security number, although the example given used only the ID number. Table lookups also avoid the need to store the same data over and over unnecessarily.

Indexes. Lookup can be expanded by indexes. Again, an illustration is the best explanation. Assume that the only information available on the employee is the ID number. A lookup table listed by employee ID number would bring back all of the needed information.

Indexes provide a rapid search tool for the user. A city directory can be listed by names or by street addresses. Name and street address are paired data that describe the same person, but the two different lists give the user the opportunity to locate the person when all the query information needed is not available. If this analogy is transferred to the occupational health and safety system, ID number and name could be the example. ID number is the key field that locates employees, but lacking the ID number, the employee can be found through the name. Name is the indexed list; by locating the name, the ID number is found.

The disadvantage of indexes is that, for every update to a file, there must be an update to every index attached to that file. On large systems this is programmed to be automatic, but on some microcomputer systems it can be time-consuming, depending on the software construction. Programmers take indexing methods into consideration when writing the system.

Two controls over index creation are the database management software and the application software. The DBMS will create the indexes from existing tables, as directed by the programmer. The developer will evaluate the indexing capacity of the DBMS, among other features, in making a selection compatible with occupational health and safety software. DBMSs do differ in their indexing approaches, which is one of the features the developer will evaluate.

The developer uses indexes judiciously to avoid slowing down the system while trying to provide the user with an efficient record searching capability. Table lookups also carry overhead, but they are used freely to save data input and avoid typing errors to preserve database accuracy.

Ways to Use Table Lookups. Across the disciplines of occupational medicine, industrial hygiene, and safety, there are commonly shared data for which accuracy is important. Locations within the plant, department numbers, and processes relate to both employees and agents, in addition to names, ID numbers, social security numbers, and job codes, plus names and ID numbers for agents. Indexes for each of these will allow users to find employees and agents by name or number, whichever is available. Agent synonyms will also need an index. Locations, departments, and processes should be evaluated for indexing needs. They will probably not require indexing, but the decision must be made by the developer.

Supervisor identities are important. Since these are employees, they carry the name, ID number, and social security number. It is not necessary to encumber the system with all three in their supervisory capacity. Name and ID number, or name and social security number will be sufficient, depending on whether the plant uses the ID number or the social security number more commonly in employee identification.

Wherever codes are used, there should be table lookups. After entering the code, the system looks up the matching description and returns it to the screen, as illustrated in Figure 3.9a. The user notes the description, which is a check on the entry accuracy, thus providing verification. This minimal expenditure of time is a form of insurance for the user. The value of codes will be discussed in Section 4.10.

Some of the data that can be standardized by coding and table lookups, in addition to those mentioned above, are shown in Figure 3.9b.

In an integrated system, some of these coded tables should be shared. For both the clinic and the safety department, body parts and the nature of injuries should be described in the same terms so that the data for the OSHA 200 and OSHA 101 are identical.

Some tables, like equipment ID number tables, may pose system

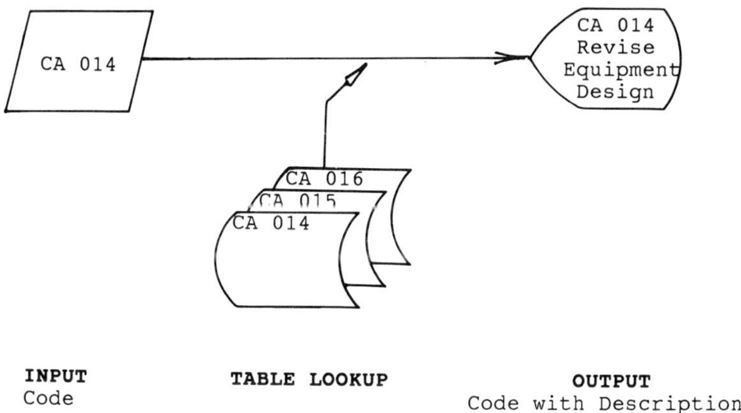

Fig. 3.9a Using a table lookup.

problems. One shared equipment table for all equipment will mean that every user has access to its contents. To preserve departmental control, it may be best to have three equipment files, one in each module. This would entail separate lookup tables as well.

Correction Factors. In a system that performs some calculations in addition to receiving data, lookup tables can also be useful. Two simple examples of this can be taken from occupational medicine: audiograms and pulmonary function tests. In audiograms, physicians often prefer to make corrections for presbycusis. Using a table lookup, the system would check the presbycusis table for the correction factor that matches the sex and age of the employee and compute the corrected values. In pulmonary function testing, the percent predicted values for forced expiratory volume at one second (FEV1), forced expiratory volume at three seconds (FEV3), and forced vital capacity (FVC) could be computed after referring to tables. Body temperature and pressure saturation (BTPS) correction values for machine temperature can also be found in lookup tables for the pulmonary function test.

One computation constantly needed throughout the system is age. Rather than enter the age of the individual on each record, the system, using a more accurate method, locates the date of birth in the demographic record and computes the age at the date of the record. This is effective in the calculations that are programmed for presbycusis corrections and pulmonary function testing. Whether the calculated age is stored permanently in the audiogram with corrected data or recalculated at each retrieval must be decided by both users and the developer.

SUGGESTED DATA FOR CODING AND TABLE LOOKUPS

General Use

Resource persons

Equipment IDs

Occupational Medicine

Body parts

Clinic case disposition

Clinic visit reasons

Diagnoses

Findings

Medications

Procedures

Protocols

Symptoms

Treatments

Industrial hygiene

Media

Laboratory methods

Reactivity

Shippers

Vendors

Safety

Incident types

Nature of injury/illness

Body parts

Source of injury

Corrective actions

Fig. 3.9b Data suitable for coding and table lookups.

3.10 WHY DATA ELEMENTS NEED TO BE LINKED

Pathways. An integrated occupational health and safety system must move between users, modules, programs, files, and variables. Pathways are needed to link together the many parts of the system, the linkages being of several types:

- Users such as the clinic and the safety department share files like injury and illness data.
- Files like laboratory blood analyses are shared by the clinic and an industrial hygienist.

- Programs such as training are shared by all three occupational health specialists.
- The MSDS module is shared by everyone on the system
- Variables such as job codes are shared by all users.

Sharing requires navigating from one part of a system to another by pathways, or links, that are established during system development. Some links can be made through 4GL system generators, while others have to be programmed directly into the software.

Types of Links. Basic types of links are:

- Variables
- Jumps
- Query paths
- Report generation

An example of variable linkage is a piece of equipment that requires calibration. Whenever an equipment ID number is entered into a record, the system automatically goes back to the record for that ID number and finds the latest calibration date; this is brought forward to the new record. This would apply to audiometers in hearing testing and to monitoring equipment in industrial hygiene sampling. Some 4GL system generators attach links to a variable when specified.

Jumps require special programming. They provide the means to shift between modules, programs, or screen forms without getting out of the screen being used. When the user finishes with the second screen, there is an automatic return to the original screen to continue data entry. This avoids the inconvenience of having to back all the way out to the original menu, go into a new menu, back out again, and return to the original screen. Programming jumps in software is almost as old as programming itself. Jumps are an original method of navigation, employed very effectively in assembly language. A long list of ways to use jumps could be given, but two suggestions may suffice to encourage readers to come up with their own ideas:

- You are deciding how much personal protective equipment (PPE) should be prescribed for an employee, and while filling out the screen for the PPE assignment, you jump to the job safety analysis for that job code to scan the record and check on all PPE required. Having made the check, you return to the original employee record and complete the PPE assignment. This is a jump within the safety module between programs (the protective equipment program and the job requirements program).

- A physician is examining an employee during a clinic visit and wants to check the employee's job assignment. The work history is brought to the screen, followed by a jump between the employee and safety modules to get to the job safety analysis for details on the weight-lifting requirements of the job. A return jump is made back to the employee work history.

These examples were purposely chosen to show jumps to the same record, but in the first illustration the jump was across programs within the same module, and in the second illustration the jump was across modules.

Query paths are navigational directions set up by query systems like SQL. The SQL system organizes the access paths between files and variables when the user logs on to SQL. The paths are like telegraph lines that send retrieval requests through the system in a fast and efficient manner.

Report generation is similar to query paths. 4GL report writers follow navigational paths to retrieve data, except that the paths are defined at the time the report variables are named. Some report requests place special requirements on the variables called *filters*. An example is to request that all exam results for cholesterol above 200 mg/dl be displayed. The filter excludes all results of 200 mg/dl or less.

The Rewards of Linkage. To the reader, linkage will appear to be a simple system task of no particular consequence to the user, and linkage between programs, modules, files, and variables will be taken for granted. However, in the occupational health and safety system, linkage has great long-range significance. For analysis and reporting, navigational paths are necessary in drawing together data from very different parts of the database. For data entry purposes, links are simply a convenience.

In the very long term, epidemiologists need to pull together related data on work histories, medical exam results, industrial hygiene exposure evaluations, and perhaps safety incident investigations. This is analogous to drawing together special data located in Miami, San Diego, Seattle, and Boston by phone conference for an evaluation of the economy. The ability to navigate completely to any part of a large, complex database is a mark of an integrated applications system.

The essence of occupational health and safety is the ability to detect low levels of hazardous exposures before injury or illness occurs in the workplace. Finely tuned recordkeeping and information retrieval are paramount to analysis of all the data required for epidemiological studies. Causal relationships can be detected through linked data from employee exposure records and medical test results, and from accident injury data and incident investigations, by measuring the degree of hazardous effect against the probable contributing factors. This is a lot to expect from an information-gathering effort. Manually, it requires extensive sifting of evidence. The real contribution of an integrated computerized system is that the data can be retrieved quickly and efficiently. The better the linkage, the faster and more efficiently this will take place. Coding of data provides standardized, meaningful terms that the computer can easily search for a match.

In helping the analyst to design a system, users should give extensive thought to the kinds of data needed for reports and analytical studies. If there is no company epidemiologist to consult when planning an occupational health and safety system, it is worth the cost to work with a consulting specialist in this field for advice on recordkeeping. Epidemiological analyses need to be considered at the time of system design. Too frequently, data are unusable in studies for lack of sufficient forethought. System design should purposely look forward to future uses and not be confined to the demands of the present. A lot of data must be accumulated before low hazard levels are likely to be detected. This means that ultimately the data will be most useful in the future. Design with an eye toward that future!

3.11 HOW EXTENSIVE CAN THE CALCULATIONS BE?

Calculations and Computations. Calculations are the mathematical algorithms used to solve a problem; they are planned methods. An algorithm is a set of rules, or group of procedures that together lead to the solution of a problem. Calculations are a human task rather than a machine task because they require reasoning.

Computations are the actual arithmetic operations that can be performed by machines like calculators and computers. Calculators are limited in performance to numerical operations, while computers can handle multiple tasks like arithmetic operations, word processing, and graphics. Computers can be programmed to perform extensive algorithms; their special usefulness is their ability to run calculations repetitively at a high rate of speed.

The public assumes that computers can do all mathematical calculations, which is certainly not true. A lot depends on the hardware and software. Complex mathematical operations may require extensive algorithmic design and programming. Some software, hardware, and programming languages are better adapted to business calculations while others are better for scientific calculations.

Calculations for Occupational Health and Safety. In occupational health and safety, there are a variety of uses for built-in calculations. Some typical types with examples are given in Figure 3.11. The list is not intended to be exhaustive, only suggestive. Complex statistics are not shown because they require special programming expertise. Statistical evaluation like regression analysis is best done with plots, and plots are difficult to program. With this limitation in mind, the developer will wisely opt to confine system design to less challenging problems and export data to another package for extensive calculations.

Screen painters usually have computed variables. These are fields defined for special arithmetic processing in which formulas are straightforward. Where more operations are required, special programming will have to be inserted into the code. An example of computations that need

TYPICAL SYSTEM CALCULATIONS

Type	How Used	Examples
Corrections	Audiograms	Presbycusis corrections
Calculations	Air monitoring	Average flow rates
		Time weighted averages
Conversions	Weights and measures	English to metric
		Metric to English
	Units	ppm to mg/m3
Counts	Reports	Incident numbers per dept.
Totals	Reports	Workmen's Compensation
Descriptive statistics	IH studies	Means, standard deviations
		Confidence limits, standard error of estimate

Fig. 3.11 Typical calculations used in an integrated occupational health and safety computer system.

more processing is audiogram corrections for prebycusis where simple arithmetic operations must be combined with table lookups.

Once computations are tested for correctness, there are several advantages to computation done by the machine. This method:

- Greatly reduces the possibility of errors in computation.
- Greatly reduces the possibility of errors in numerical data entry.
- Ensures accuracy when repetitive computations occur.

Raw Data and Corrected Data. A question that is invariably asked by developers is, what data need to be stored and what data only need to be displayed? Computations can be processed either upon entering raw data or upon retrieval. If the computed data have special importance, it is best to store them and make them an integral part of the record, along with the raw data. Always store raw data, but if the computed data are only transitory in nature, carry out the computations at screen display. A simple example is the average flow rate of a pump during air monitoring, which is computed from the initial and final recorded flow rates. The average rate can be stored along with the initial and final readings, or it can be computed each time the readings are brought to the screen. The developer will be guided by the user's preference.

Decisions like this one are based on the requirements of the professional users of the system. The systems analyst will explore each kind of computed data with users, and may balance user preferences against machine processing burdens and storage capacity before helping the users to reach a final decision. Storage capacity on a large machine will not present problems, but a microcomputer may need to conserve storage space.

Statistical Computation. Descriptive statistics are readily handled in DBMSs, applications software, and most 4GL generators. Operations like arithmetic means (averages) and standard deviations are commonplace. Some may offer 95% confidence limits and a standard error of the estimate. With the addition of regression analysis, programming is elevated to a higher level.

To avoid complex programming, include only simple descriptive statistics, provide the user with the ability to subset data for further statistical evaluation, and export the subsets to packages like SAS and SPSS for more rigorous statistical operations. Microcomputer users are wedded to their Lotus 1-2-3 spreadsheets, which serve their statistical needs admirably. Consequently, it is more sensible to avoid reinventing the wheel and to design the occupational health and safety system to include only simple statistical operations, providing a programmed interface in order to export files to other software for further analysis.

Industrial Hygiene Calculations. It will be useful to design specific computations for industrial hygienists. Most computers can handle logarithms, making it easy to program geometric means. Two standard deviations will then be used, one for the arithmetic mean and one for the geometric mean. Programming can be inserted into the code for running confidence limits. While most hygienists consider only 95% limits to be necessary, some may require other selections.

Special-Purpose Modules. Numerous special-purpose microcomputer packages are available in occupational health and safety. Hearing conservation programs and MSDSs are easily included in an integrated system. Special packages that perform specific tasks like ventilation modeling and plume dispersion modeling are stand-alone systems. They require the graphics capabilities provided by underlying graphics software. Graphics to go with calculations and reports may be desirable from the user's point of view, but it is wiser to subset the data to Lotus or special packages rather than to integrate them with the occupational health and safety application.

The purpose of an integrated occupational health and safety applications package is to integrate data input and output for the health and safety disciplines, which is not the same as integrating every kind of computer use into a single package. The software underlying the applications package is specialized for database management. A programmer can make it easier for the user to switch between packages by building a front-end menu that allows the user to select access to special-purpose packages or to the occupational health and safety modules.

Porting Data. To port data from the occupational health and safety system to statistical and spreadsheet packages, select those fields which are important to process rather than attempting to port already processed data. Within the occupational health and safety system, use descriptive statistics on medical test data, for instance, and when the results look interesting, port the original data to a statistical package for further study. Use descriptive statistics for trial and error to find an appropriate combination of data to test rigorously. Then subset those fields and port them across the interface. The new package will recompute the calculations and store the results in its own database for additional analyses.

It is easiest to retrieve and port data that have been coded. Text data are useless for arithmetic operations. Be sure that data are coded so they are ported for analysis, the epidemiologist can be confident that apples are being compared to apples, not oranges.

Data Analysis. Analysis is what follows data retrieval. This requires data manipulation, which differs from the processing of data during input and output (retrieval). The DBMS is designed to make files, input records (data), store them, and retrieve the records from files.

The next step is to look at different combinations of the data. This could be called *ad hoc reporting* and it is an effective way to subset data. When a combination is found acceptable, subset it and port it for statistical analysis. If the results are unsatisfactory, return to the database application and sample additional combinations. The output of the statistical package will lead the user to final conclusions and a study report.

3.12 WHAT WILL IT TAKE TO INTERFACE WHERE NECESSARY?

Interfacing. An interface has already been described as the boundary between two different environments. This presents technical problems in the computer world because there are different kinds of processing chips, different kinds of equipment, and different kinds of electronic signals. In computers, interfacing is a procedure that adapts the signals flowing from one device to the signals used by a second device. Even within chips, signals have to be adapted for different functions, like input and output, which differ from memory operations.

The kinds of interfacing the developer will investigate are external to the computer system. Typical examples are sending or receiving signals between the computer and:

- Data loggers and technical instruments.
- Other database systems like payroll and personnel.
- Communications systems like networks, telephone lines, shared databases, and other computers.
- Peripheral equipment like printers, plotters, and tape drives.

Interfacing to Data Loggers and Technical Instruments. Data loggers and technical instruments normally operate on analog signals, while computers operate on digital signals. Analog systems represent data by recording changes in a physical condition like voltage or pressure. Digital signals represent data by bits (digits like zero and one) or groups of bits (bytes). Signals passed from instruments to computers have to be converted from analog to digital, for which there are analog to digital (A/D) converters.

For example, spirometers and industrial hygiene monitors require A/D conversion. A tracing is made by the spirometer that records changes in pressure. This analog signal is converted to an electrical pulse by a part called a *transducer* which, in turn, is converted to a digital signal by an A/D converter. The signal is now prepared for reception by the computer.

Data loggers have computer chip parts. They process the data received and often print out the results on paper tape. Instruments like spirometers are increasingly adding computer chips and small printers, providing users with convenient printed output. Chemical-detecting instruments set for timed results take measurements at intervals of a second or more for lengthy periods, record all the data, perform computations, and produce results integrated over a period of time. Some of these instruments have computer interfaces so that the data can be ported to floppy diskettes in a microcomputer. These data could be ported to an occupational health and safety system with the proper interface.

Interfacing to Other Database Systems. A primary necessity in the occupational health and safety system is to receive demographic and job history data from the personnel and/or payroll departments. For the sake of simplicity, it will be assumed that the transfer takes place on the same computer, with no need for communications devices, tape drives, or floppy diskettes. The only environments to interface are two different DBMSs whose file formats may be different. A small program can be inserted into the first database to select a file (e.g., personnel) and to select the required fields. The fields will need to be sequenced to match those in the demographics file, requiring sorting of the fields. Field lengths may also need to be matched. The program will organize the personnel data to match those in the demographics file. A second short program that is compatible with the occupational health and safety software will stream the data into the demographics file. These are not complicated programs, but they do require additional programming.

ASCII files. Transfer between files (on different computers) is usually done by an intermediate file that is readable on both sides of the interface. The typical file code is the American Standard Code for Information Interchange (ASCII). It is a standard for the communications industry. There are other file transfer codes, but ASCII has become the norm and serves to illustrate file transfer adequately. ASCII is a 7-bit code, whereas the word length for computers is 8 bits. Seven bits (combinations of 1s and 0s) comprise the ASCII representation of a character (A, y, 2, $, etc.), and the eighth bit is used to check for errors in the transfer stream. This is called the *parity bit*. If the transfer requirements are for even parity, the sum of the bits in the byte must be even. The check is made after transmission. If the sum is not even, an error is assumed and the data have to be resubmitted for transmission.

ASCII provides for 128 characters that include upper- and lower-case letters, numerical digits, graphic characters (like the comma, period, question mark, and exclamation point), operators (like +, −, <, >, and =), control characters (like a carriage return, tab, line feed, and printer controls), and foreign language letters (like cedilla, circumflex, and umlaut). This is a very flexible transfer medium, and tape drives are used to transfer data between computers (microcomputers, minicomputers, and mainframes) because they operate in ASCII.

IBM has its own file transfer code called the Extended Binary Coded Decimal Interchage Code (EBCDIC), but this is not an American standard, as is ASCII. The bit arrangement differs, and in those cases where file transfers have to be interfaced, further conversion has to take place.

Interfacing with Communications Systems. Data transfer across communications systems is usually done through ASCII, but it can also be done by some other system, like binary code. The device that interfaces between the computer and a telephone line is the modem. The term *modem* is a contraction of *modulator-demodulator*. Modems have their own software to set up the mechanics of data transfer. For transmission, the modem encodes signals compatible with fiber optics, copper wires, microwave dishes, and satellites. The receiving modem decodes for computer use. These communications devices transmit signals as frequency tones. The binary digit 0 is a low-pitched tone, and binary digit 1 is a high-pitched tone. Thus, with the aid of modems and communications lines, ASCII data can be encoded for transmission to most places on earth and decoded upon receipt.

A communications capacity enables users to tap into shared databases in other geographic locations, which may be corporate databases in a distant headquarters or public databases for technical information. An example of a useful public database is a reference database that OSHA has developed on hazardous workplace conditions. This is a service provided to industrial hygienists and safety professionals throughout the United States that began as a staff database for OSHA. It has become a resource system of about 235,000 records and a collection of a dozen smaller databases. Information is available on chemical profiles, interpretations of OSHA standards, reference bibliographies, sample analyses, and training materials, among other important subjects.

Time-sharing on remote occupational health and safety databases is also possible through modems for the small company that cannot afford the cost of machinery and applications software for automated recordkeeping. This may also be useful on a temporary basis while awaiting the completion of software development.

Interfacing with communications systems makes possible decentralization of a corporate system and networking, as discussed in Section 1.4. Still another use is transmitting data from one computer to another from medical laboratories and analytical chemistry laboratories into blood chemistry and other medical module laboratory tests, as well as the sampling analysis results in the industrial hygiene module. Some labs transmit by telephone to a modem with a built-in digital readout. With programming at each end of the communications system, it is possible to run the data directly into computer files. This has been under experimentation and is a form of batch processing for which special interface programming is required.

Interfacing to Peripheral Equipment. The type of interfacing required for peripheral equipment requires some device setting within both the computer and the equipment, but the chief technical interface problem is connecting cables to send the right signals. Computer settings are done by tiny switches.

The computer has ports (gateways) for attaching cables that go to equipment like display monitors, printers, plotters, and modems. Many directions specify that an RS-232C cable connector be used. The Electronic Industries Association (EIA) has standardized connectors, so that they are universally adaptable in the United States to ensure compatibility between data transmitters and computer equipment. The standard connector is the RS-232C which has 25 pins inside the male connector, each with a specific function. To control load capacitance on the line, the cable length is restricted to 50 feet (15 meters). For distances over 50 feet, other means of carrying signals, like telephone lines and fiber optics, must be used.

Computers operate on parallel data transmission between chips, meaning that all bits in a byte are transmitted simultaneously. Phone lines and some peripheral equipment operate on serial data transmission by sending the bits in each byte serially (one after the other) along the transmission line. Thus, the trick is to convert the serial transmission to parallel so that the computer can send the data to memory or storage.

A small chip called a *universal synchronous receiver transmitter* (UART) does the actual conversion from parallel to serial and from serial to parallel, since it has separate input and output function areas. Operations in both directions can be carried on simultaneously. UARTs are to be found in computers, modems, and data loggers, making possible the conversions between serial and parallel transmission. The UART is a key device in an interface.

The RS-232C connector and the UART together constitute the physical interface between computers and equipment in the United States. Other countries have their own interface standards set by the Comité Consultatif International de Télégraphie et Téléphonie (CCITT), making it necessary to modify U. S. equipment for use in other countries.

3.13 SCREEN DESIGN

The User Interface. The most important interface for the analyst to design is that between the user and the machine. The keyboard comes with the computer, but screen display is a part of software development and is a critical component of the user interface. This is a product that demands close cooperation between user and analyst in the development process. Screen display and its relationship to paper forms will be discussed in Sections 4.2 and 4.3. In this section, the emphasis will be on how the developer can adapt the screen display to the user's advantage.

The conditions that make a user interface necessary are shown in Figure 3.13a.

THE USER INTERFACE IS:

What a user sees

What an application program provides

What appears on the video display screen

What initiates interaction between the computer and the user

What initiates responses between computer and user

What makes a complex system transparent to the user

Fig. 3.13a Reasons for having a user interface.

User Interface Design. Screens are less readable than the printed page. Reading a screen is a slower process than a page. Therefore, the screen must be designed for readability (comprehension). The resolution of a VDT is limited by the number of activated phosphors on the screen. Technology to increase tube resolution by narrowing the space between phosphors is improving all the time. The average computer terminal screen is not on a par with the high-resolution screens now available for very good graphics; for this reason, alphabetic and numeric characters suffer in legibility. Screen resolution does not equal that of print on a page, leaving it to the developer to compensate with features like enhancements (discussed in Section 3.6) to improve use. Newer screens generally have higher resolution.

If the screen is properly designed, users should not have to rely heavily on documentation. Users want to concentrate on their work, and disruptions provoked by the system that force them to consult reference manuals are resented. Users prefer minimal reading, and when the added frustration of inadequate documentation occurs, systems can be abandoned as "not worth the trouble". Consequently, the developer has a responsibility to design a screen that promotes easy use of the system.

This begins with easy-to-use menus attractively designed for swift item recognition and selection. Instructions for using the menu screen beyond the initial training directions should be unnecessary. There should be no ifs, ands, or buts to a menu screen. Selection should be direct, with a single keystroke or mouse click, leaving the computer to do the work of taking users to where they want to go.

The User-Friendly Spirit. The user-friendly features that are designed into the system should be transparent to the user. This can be accomplished by:

- Using the language of the user.
- Avoiding abbreviations unfamiliar to the user.
- Being consistent across all screens.
- Creating a logical sequence of data on the screens.
- Creating a logical sequence of screens.

These features enhance user comprehension of the data. Too little data on the screen disrupts comprehension, and a crowded screen slows reading speed. When similar data are grouped, reading comprehension is reinforced. To improve comprehension have users help to write the prompts and screen titles, thus ensuring maximum user/system cooperation.

Data entry is a time-consuming task, and therefore expensive. Screen design must maximize entry efficiency and minimize user effort. Single or combination keystrokes help to speed up entry functions. Word processors have so many functions that they rely heavily on keystroke combinations and function keys. Spreadsheet users can invent macros to increase efficiency. In reality, however, only some dedicated users exercise all these options; most users simply memorize only the necessary ones. A lesson to be learned from this is that the analyst and programmers can design a lot of remarkable little tricks on the computer, but the user is not likely to appreciate them. Users learn to use keystroke combinations and function keys only with practice. Anything they do not practice in the natural conduct of their work is best left uninvented.

Problems to Solve. Occupational health and safety presents some problems that are soluble but that challenge the abilities of both analysts and programmers. The first problem is how to deal with text. Users are so accustomed to filling in forms with textual responses that are extended to the back of the page that they will be reluctant to learn that there are better methods of entering data. Codification of data is the best solution to the problem and is discussed in detail in Section 4.10. The developer will be diligent in determining what text is unavoidable and how to handle it in the computer system. While coding is the key to elimination of a lot of text, some text can be a valuable part of the database.

Text entry presents two problems for the system designer. The first is the amount of space to allocate in the database. Some DBMSs require specification of total file lengths, and data dictionaries require specification of field lengths. Text entry is a mixed bag of unpredictable line lengths. If the designer allows 300 characters for a field and the majority of users use only 250 characters, after 1000 records this field has encumbered 50,000 characters of storage space that will never be used. With more variables like this one, available storage space will soon be exhausted; this is an acute problem for microcomputer hard disks. The solution is to limit text to those variables that cannot be codified or answered in a maximum of two or three words. The developer can provide special programming for the variable that stores the text outside of the file. This requires the system to record the storage location where that text resides, and a pointer has to attach the database variable to this location so that text can be retrieved when desired. This solution will allow unlimited text when required.

The second problem in occupational health and safety is the unpredictability of the number of entries needed for some variables. The developer would have to decide on an optimal number of entries, which is an impossible task. Examples of such variables are the number of items of

PPE assigned to an employee, the number of conditions or acts that contributed to an accident, or the number of separate injuries found when someone is taken to the plant clinic. Like the problems with text, solutions will vary with the system and its developers, but solutions are possible, challenging analysts and programmers. Specific solutions cannot be offered here because they are proprietary to developers and may be copyrighted. There is no generic solution.

In both text and multiple entries for a single variable, the problems of retrieval must be kept in mind. How do you find something embedded in text without going through time-consuming string searches? String searches are word sequences for which the system will laboriously search for an exact match. Consider the problems when one body part is described by different people as:

Arm, right
Right arm
Right forearm
Forearm, right
Arm below elbow, right
Arm above wrist, right

The computer, being robotic, cannot be efficiently programmed to search for every combination. In addition, there will always be the person who writes instead *right ulnar,* or *right radius,* and the record is forever lost to retrieval. The solution is codification which everyone is obliged to follow for the sake of consistency.

Scrolling. Screens are typically 80 characters wide and 25 lines long. Many occupational health and safety forms are two or three times that length. One solution is to break up one long form into two or more smaller forms. Another is to scroll long screens up and down. Scrolling is a typical aid to getting many lines on one screen.

Figure 3.13b is an example of a demographic screen form that exceeds the 25-line limit. The screen will initially show all lines above Personal Physician. As the user enters data and reaches the last line in the Emergency Contact section, the screen will scroll up to reveal the remainder of the record.

In many screen painters, there is a one screen/one record ratio, which is a handicap. It is better to match the length of the screen to the record requirements than to match the record to screen limitations.

Scrolling moves a block of lines up all at one time. Paging shifts an entire page up as you reach the bottom line. Scrolling is a user-friendly

```
MODULE: Employee                                           PROGRAM: Demogr
                         EMPLOYEE DEMOGRAPHIC DATA

EMPLOYEE ID: 9876                         NAME: Benjamin A. Levy
SOC.SEC.NO.: 333-44-5566                   AKA: Benny
-----------------------------------------------------------------
    BIRTH DATE: 11/30/1965          HIRE DATE: 09/15/1985
          RACE:  C White          EMP. STATUS: FT Full time
           SEX:  M Male            TERM. DATE:
MARITAL STATUS:  S Single        TERM. STATUS:
 SPOUSE'S NAME:  None

HOME:
  Address 141 Maple Ave.              Phone No. (H) (987)456-6677
          Apt. 4-B                    Phone No. (W) (987)654-7800
     City Anytown                                Ext 63_
    State MA     Zip Code 34567

EMERGENCY CONTACT:
     Name Rosa Levy                   Relationship  Mother
  Address 220 Elm Ave.                Phone No. (H) (987)457-5678
          ______________________      Phone No. (W) ____________
     City Anytown                                Ext ___
    State MA     Zip Code 34567

PERSONAL PHYSICIAN:
     Name Richard Emerson             Phone No. (987)457-0987
  Address 130 Oak Ave.                          Ext. ___
          ______________________
     City Anytown
    State MA     Zip Code 34567
```

Fig. 3.13b Features of a user-friendly interface.

feature that always leaves a few previous lines in view so that the user does not lose relevant information. It has definite advantages for screen forms that exceed 25 lines.

User-Friendly Screen Features. There are many user-friendly features that the developer can design into the screen, such as those shown in Figure 3.13c. In addition to these, give the user help with the fields. Highlight, underscore, or by some method outline the length of the field. If there is a one-character response like "Yes/No," indicate what is needed and do not expect the user to guess at the requirement. Any prompt that requires a "Y" or "N" should be followed by "(Y/N)" for instant recognition. The same is true for any other abbreviated responses to prompts.

Users will have varying experience with computers. Screens have to be designed down to the level of least experience. More experienced users will ignore many of the screen features so necessary for the inexperienced user. The important objective is to encourage inexperienced users to

USER-FRIENDLY INTERFACE FEATURES

Easy logon and logoff directions

Attractive screen presentation

Easy to read screen presentation

Well-proportioned screen composition

Open space to avoid crowded look

Headings and subheadings over
grouped data

Fast response time

Default values wherever possible

Delimiters wherever possible

Multiple windows permissible on a
screen

Fig. 3.13c User-friendly interface features.

become experienced by providing transparent screen interface methods that make them believe that the system is user friendly.

3.14 A DATA DICTIONARY

Origin of the Data Dictionary. Structured program design proposals in the late 1950s led to top-down programming and culminated in the 1960 announcement of a new language called the COmmon Business Oriented Language (COBOL). It was organized in four major divisions: Identification, Environment, Data, and Procedure. The first two simply described the program and its environment. The Data Division defined the files to be used and preceded the Procedures Division, which contained the actual program commands for processing the data. The data dictionary evolved from the Data Division.

The COBOL divisions could be divided into sections. The Data Division contained a file section that described each file in detail, the file name, the maximum record length (up to 80 characters), and the maximum number of records expected in the file. Fields were named and described in detail in terms of their type (alphanumeric or numeric), their length, and their sequence in the file. Even carriage controls and spaces between words had to be accounted for. The limit of 80 characters in a file matched the early 80-column keypunch cards. Needless to say, a file contained few lengthy fields, which was a compelling reason for data codification. Finally, the files had to be designated for input or output. COBOL is a very thorough programming language, but it consumes so much memory that it has not been a language for microcomputers or minicomputers. The principal use of COBOL has been for IBM mainframes.

Since the mid-1970s, data dictionaries have come into use. Originally, data dictionary systems were commercially available. A dictionary system not only has a dictionary, but also a command language and a screen interface, like other programmed applications software. It is a stand-alone system and accesses its own dictionary. Data dictionary systems evolved into database systems and then DBMSs, carrying along the original COBOL data definition concepts. A DBMS is not required to have a data dictionary, but most do.

The data dictionary provided a structured method for creating a database that not only defines the database during the logical planning phase of development, but that can be incorporated into the physical system and continue to serve a useful purpose for the life of the software.

A Catalog of Data. As the database is being created, a large quantity of detail emerges that quickly gets out of control. There are suddenly dozens of data items to be remembered and each one has its own features, multiplying the working detail several times over. The demand for an organized catalog of data is immediately apparent. This catalog helps the user to keep track of what is needed for records, and it helps the developer to eliminate redundance from the database.

The data dictionary is more of a catalog than a dictionary. It describes the data and may give meaning to the system name for the data element. As in a mail order catalog, every item has an identifier. Rather than a catalog number, the identifiers are coded system names for data, not the user's normal name. Most systems limit the identifier to eight characters. Each one has to be unique, and trying to name every data item becomes a headache. It is like trying to give every child in a school of several thousand population a unique name. Fortunately, there are 26 letters and 10 digits to work with, yielding a lot of unpronounceable arrangements. (1,263,618,323,136 permutations, to be exact, if the characters are taken from one to eight at a time. This happens to be a little less than the national debt!)

To control database growth, it is advisable to have naming conventions, like all dates starting or ending in -DAT or all names starting or ending in -NAM. This helps users to recognize the cryptic code name for the field or to determine what it stands for. When there are measurements like frequency readings in audiograms, the names can include numbers. Since the name does not have to use all eight characters, it can read FREQ8K or AUD8K. Where series of things occur, names like PROCESS1 or PROCESS2 can keep the fields grouped together in the catalog. The important thing is that responsible programmers make system names understandable, knowing that some day the user or other programmers will need to use them.

Describing Data. A description of any object includes measurements of its size, a category to which it belongs, a picture of its appearance, and an indication of its location or its relationships to other objects. Data are categorized by their types. Alphanumeric data can include both letter and numeric characters. Numeric data can only have numbers but can be specified as integer (whole) data or real data that contain a decimal point in a specified position. Data sizes are the number of characters in the length of the field. Length is part of the picture, along with the type of data. The location is a specified position in a given file.

In addition, the data may have some special importance, such as being a key field. Although each file bears a unique name, records also have to be unique in order to be found in a file. The identifiers are key fields. Files may need more than one key. To illustrate, the PPE assignment file may be keyed on the employee to which the piece of equipment was assigned. Many employees may have been assigned more than one piece of equipment. This necessitates a second key that designates which type of equipment was assigned to the employee. The employee ID number will be the first part of a key, and the equipment type (equipment ID number) will be the second part of the key.

The Dictionary as a User's Guide. During development the data dictionary will be continually referenced by the analyst, programmers, and database manager. It may be some time, however, before users become dictionary conscious. If users avail themselves of facilities like SQL and a report writer, the data dictionary will be useful for a query and report structure or for subsetting data for export to another database. Users will eventually discover the dictionary and will make the effort to learn system names. Through good training, users learn to turn to the dictionary for help like finding key words. Use of the data dictionary can create knowledgeable users.

For SQL users the data dictionary can be queried just like the database, making it a valuable tool for locating file and field names or having data descriptions readily at hand. Printouts of data dictionary information are occasionally needed, and this is a fast means of obtaining them.

Throughout data entry or modification, the system will use the data dictionary for data consistency checks to preserve database integrity. Thus, the data dictionary becomes the unrecognized guarantor of user confidence in system dependability. It also guarantees future reliability during maintenance and repair. When connected to log files, a historical record of changes in the dictionary will contribute to the epidemiologist's reliance on the database.

In a distributed system or network, where many users share data, the data dictionary is a vital part of the system that requires all users to refer to fields by their unique names, a feature that ensures data integrity across several or more workstations.

The user should understand the distinctions between the data dictionary and a table lookup. The data dictionary is generated by the DBMS or the application generator, while the tables for lookup are generated by the users following implementation. The data dictionary only verifies that the data being entered meet the original data specifications for that field; it contributes nothing to the screen display or the stored record (see Figure 3.14). The table lookup supplies the description or definition of an input code. The definition is not stored in the record; only the code is stored to save storage space. Each time the code is brought into memory, the table lookup supplies the definition to match.

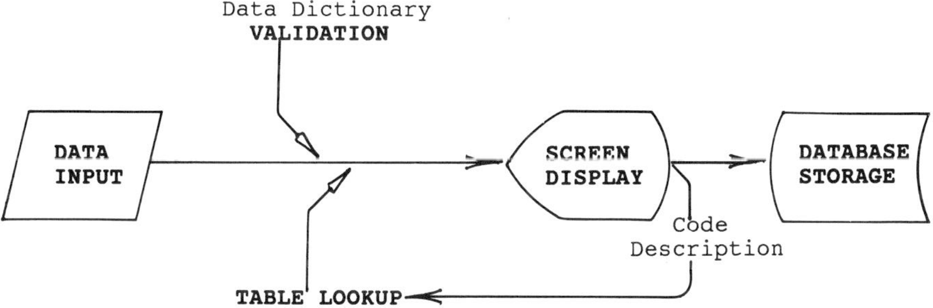

Fig. 3.14 The distinction between a data dictionary and a table lookup.

3.15 SUMMARY

Vendors of integrated occupational health and safety software are very willing to demonstrate their products to users and information services for consideration, but the final decision may be to custom develop the product with a contractor or in-house information services. It is more costly to develop a system than to purchase it, but the cost can be reduced by using 4GL tools like screen painters and report generators. The developer should be chosen for business reliability, experience in developing application software, software experience, and software management ability.

Applications development is greatly enhanced by prototyping, which is one of the newer methods of software engineering. The repeated cycles of design and user review in prototyping lead to a product that is specific to the user's needs. The system design should represent the joint efforts of a systems analyst and the users.

Software design is an engineering process that can be done by several different methods. Top-down structured design is the usual choice. CASE tools may also be used for cost-effective design, and some are integrated with screen and menu-formatting procedures.

Making the system menu driven will render it more functional so that it relates to the user's work practices and provides user-friendly access to any area of the system. An interactive human-computer system furnishes feedback to the user on what is happening within the machine. The screen is designed to keep the user informed about what the system is doing.

Physical security is the prevention of physical destruction of hardware, software, and the database. System security is the prevention of incorrect data or loss of data because of system failure or human error. Privacy and confidentiality are to be protected as much as possible in a computer system. A security administrator oversees security by establishing policies and procedures. A variety of precautions can be used to ensure security.

Database integrity can be protected by making sure that data entering the system are consistent, valid, and verified by comparison with the data dictionary. In the event of system failure, there are methods for recovering the database.

Table lookups are a deterrent to data entry error and save entry time and energy. Indexes increase data retrieval efficiency. Linking screens and programs give the user efficient ways to navigate through the system. Calculations can be attached to variables as part of the variable processing procedure, or they can be inserted into the system for required computation. Data can be stored either in raw form or after numerical transformation. Descriptive statistics are appropriate to the occupational health and safety system, but data should be exported to statistical packages for more rigorous evaluation.

Software interfaces can be programmed between the occupational health and safety product and specialized packages. Data are usually transmitted by ASCII files, a method which is also used in communications systems. Peripheral devices like printers, plotters, and modems are interfaced to the computer by hardware methods, which are cables.

Screens are user interfaces that are designed to communicate computer actions and conditions to the users in a user-friendly manner.

A data dictionary is a useful tool that keeps track of information about the field. It can be used to check data consistency and to define the database for the user.

Chapter 4

CONTRIBUTIONS TO THE SYSTEM BY THE USER

Occupational health and safety is still an evolving field and many of its professionals are still grappling with their roles in relation to company management, especially in industrial hygiene and safety. A repository of data is a management tool for information needed when decisions are to be made. Paper files are being turned into electronic files because computerized information retrieval is much faster and more cost-effective. Our present dependence on information demands that our recordkeeping be accurate and dependable.

Users provide the developer with the data requirements for system and database design. The primary sources for data specifications are paper forms already in use. As computer system design progresses, it may be necessary to revise some forms to improve the manner in which data are recorded. Data specification helps to describe how the data should look upon entry into the computer. MSDSs have a variety of specifications that require special design attention. Data should be reviewed carefully for potential coding, an important step in standardizing information for improved and efficient data retrieval.

System testing should be done by both developers and users. Testing, training, and implementation take place concurrently, which makes this period of development a busy time.

4.1 SELECTING THE RIGHT DATA

The Professionals. In a joint effort to develop an occupational health and safety system, the team consists of an analyst and programmers who contribute their software applications knowledge and users who contribute their specialized technical knowledge in occupational medicine, industrial hygiene, and industrial safety. This is an alliance of impressive talent that can generate a very sophisticated and sorely needed computer data collection and retrieval system. Computer technology is pervasive in our lives: in offices, on assembly lines, at all levels of education, in laboratories, in transportation, in stock markets, and in communications, for instance. Given these circumstances, the wonder is that it has taken so long to automate occupational health and safety. A partial reason is that the technical experts are still striving to define their roles and may find it difficult to decide what the system should represent.

It is not that occupational health and safety is a new concept world; it is that, unlike the practice of medicine, unlike the laboratory analysis of chemicals, and unlike the operation of production lines, new responsibilities keep being added and experience is being acquired. This is more true in industrial hygiene and safety than in occupational medicine. An anomaly exists whereby there is an urgent, immediate demand for specially trained professionals at the same time that the professionals are still trying to define what they are to be—professional advisors, business managers, or both.

Defining the Professional Role. Safety managers and industrial hygienists have had the most difficulty in defining their roles. With extensive assistance from their professional organizations (the American Society of Safety Engineers and the American Industrial Hygiene Association), they are making progress. Safety and industrial hygiene professionals are usually appointed from other areas of the plant. They are instructed to assess risks, to control hazards, and to contain exposure and accident costs, all on a limited budget. In addition to trying to perform a highly technical job, they discover that they are operating in a political climate complicated by government regulations, threats of litigation, and competition for budget money. Through professional development courses and certification, professional organizations help their members to master the technical aspects of safety and industrial hygiene, but there is no easy way to learn to juggle technical expertise and management demands all at once.

The missing link is the development of skills in risk management and business management. While risk management is a field by itself, business management covers an assortment of fields like production, planning, fiscal matters, personnel, and marketing. The company exists under pressure from government, legal advisors, stockholders, competitors, and unions, and professionals no longer find themselves in the protected environment of the analytical chemistry lab, the engineering department, or the doctor's office. It is a highly geared blend of management and technical responsibility that is political and competitive. Decisions on technical issues are unavoidably colored by management considerations for which the hygienist and safety manager are typically unprepared. College curricula in chemistry and engineering do not include courses in industrial and risk management.

Expert Data. This dichotomy of job demands is significant to the development of an occupational health and safety system. During development, professionals are the guides identifying the data that must make up the database and specifying how the system design should look. They have to decide what the objective will be: Will this be a repository of technical data, or will it also be a management tool? If the system is limited to technical data, problems will arise from the necessity to survive in a business management atmosphere. If both technical data and business data are included, what business data need to be added to round out the technical data?

Because system design starts from square one, it provides an opportunity to incorporate new data needs into the occupational health and safety setting. To add business data, there should be design input from risk management, purchasing, accounting, and anyone else whose work relates to occupational health and safety, while avoiding any indication that the new system will infringe on the turf of others. The inclusion of business data will pay off when reporting to management and preparing for the annual budget battle.

Occupational health and safety professionals have to speak in two languages, "technicalese" and "managementese," the former in their professional role, and the latter in their business role. It is a double life that is joined in the database.

Occupational Medicine. Physicians and nurses in the occupational health and safety field seem more at home with business management, possibly because their experiences in private and hospital practice take management and accounting matters in stride. Hospital budgets are notoriously limited, and the medical world has a management hierarchy of its own. The clinic tends to be an adjunct of plant business, not a part of it. Clinic operation is not dependent on production quotas, sales quotas, and what the competitors are doing. The effort to deal with clinic budgets does not seem as formidable as with safety and industrial hygiene.

Yet, in the occupational health and safety setting the clinic needs to closely cooperate with safety and industrial hygiene. Reports to management must reflect joint objectives. Most medical professionals keep records on costs and services which will be included in the occupational health and safety system.

Some Objectives in Database Design. The primary objective of the occupational health and safety system is to make certain that all necessary data are included that relate to measurements on hazardous materials and conditions that may affect employees and/or the community in any manner. A secondary objective could be inclusion of all data that assist occupational health and safety professionals in providing reports and presentations to management. Another objective is to help occupational health and safety departments prioritize their responsibilities. Which employees and conditions require more surveillance than others? Which conditions need modification to make a safer workplace? Which problems need constant attention? Which ones wait for a period of time? For how long? Which materials or conditions constitute the most hazardous exposures? How hazardous are they?

After hazardous materials and conditions are prioritized, a second level of priorities must be established for competing priorities. What risks exist if Priority A receives more attention than Priority B? This is the level where risk management knowledge will be helpful. Risk assessment will permit the professional to make an incisive presentation to management when additional funds are sought.

If the database includes data on costs and savings, presentations to management will be more forceful. It is better that occupational health and safety personnel present their own costs and savings data than to depend on support from payroll, accounting, and the comptroller, whose data will have a different emphasis.

Prioritization, risk assessment, and cost analysis are business arguments that management can understand. They are secondary to the primary objective of occupational health and safety, which is to monitor and safeguard against hazardous materials and conditions, but they can furnish compelling support for proposals related to control of hazardous exposures. The real expertise in keeping occupational health and safety data is proved not by how many records you preserve, but how well you can use those records to make a case. The case may be one made to an employee, another professional, a union, a court, the government, the CEO, or the board of directors. The excellence of its presentation rests on the kinds of records and data used. When the evaluation of technical data reveals a need, management can best relate to that need if it is expressed in business terms and supported by business-type data.

Epidemiological Data. An article written by Pamela S. Kerr and published in the March 1978 issue of the *Journal of Occupational Medicine* should be required reading for every person in occupational health and safety. It is as true now as it was when published. While many improvements have taken place in occupational medicine, industrial hygiene, and safety management, a basic problem of the last quarter century, so well described by Kerr, still exists. We lack the "sound epidemiological data necessary to correlate disease with occupation," and what we do have is either inaccurate or insufficient to validate reliable conclusions.

Government Data. Data evaluation on a national scale has been limited to the Bureau of Labor Statistics (BLS) and illness/injury counts in segments of industry categorized by Standard Industrial Classification (SIC) codes. These statistics point to hazardous industries but tell us nothing about the underlying causes and results of occupational illness. All we have are studies of small populations of workers in specific settings reported in professional journals. What we lack is data on large populations that will provide a sufficient sampling base to yield statistically valid conclusions.

A case in point is the asbestos problem. We have abundant evidence that asbestos fibers are carcinogenic, some varieties being worse than others. Although we have ample evidence from both human and animal studies, the question remains as to what is a dangerous exposure for humans. It may be that 1 fiber under one condition (e.g., heavy cigarette smoking) is as dangerous as 100 fibers under another condition (no smoking). Wide disagreement precludes any knowledgeable estimate of what maximum or optimal levels of asbestos exposure humans can endure without a carcinogenic risk.

How do we obtain valid data in occupational health and safety? Two requirements are fundamental. The first is a database compatible with interdisciplinary input, and the second is a large enough sampling population to validate the statistical evaluation. We do not judge an industry (SIC code) injury rate on the results of a few companies. The BLS collects data from all companies in that industry and combines them to obtain an overall conclusion about industry, despite differences in geographical location, environment, plant size, and plant operations. The breakdowns by OSHA have now been outgrown. We need more definitive divisions for reporting, and we need direct relationships between agent and disease conditions.

The government initiative is in the right direction. The actions now need refinement, an uneasy process in a large bureaucratic environment. What OSHA and the BLS have accomplished is to target high-incidence industries as a beginning step. Professional organizations have cooperated by offering some sage advice.

EPA has only begun an arduous task with Title III reports to pinpoint the extent of environmental change brought about by emission of hazardous materials into the soil, water, and air. The first step is to locate the agents and determine their quantities. Further steps must be taken to begin to correlate Title III data statistically with environmental alterations, and eventually leading to damage control.

Government information requirements are only in their infancy, and more intensive reporting demands are bound to come from both OSHA and EPA. If recordkeeping has been a demon in industry's midst, it is time for industry to meet the challenge with a positive attitude, because recordkeeping demands will not go away. Kerr wisely counsels that "ill-advised legislation has come about largely due to the lack of good data or the absence of any data whatsoever." When industry joins forces with the government, a consensus can be achieved about what data are needed and how to use them. Data standards are bound to expand beyond the present OSHA 200 and Title III reports. Nationwide collection will bring about electronic input, categorized not only by industry but also by hazardous agents, job descriptions, and hazardous conditions. For the occupational health and safety professional, this will have a profound impact in dealing with top management because there will be compelling evidence for the need to invest money in surveillance and controls. A proactive occupational health and safety system will become a reality.

4.2 PAPER FILES AND ELECTRONIC FILES

Required Records. As has already been pointed out, keeping records is a normal function of investigation and inquiry. In occupational health and safety there are government requirements for recordkeeping, as promulgated by OSHA, EPA, the Department of Transportation (DOT), and other agencies. Specific records, how long they are to be maintained, and their accessibility for inspection by government representatives, employees, and the public have been described. Agencies also require certain reports, which in a real sense is a recordkeeping mandate, because you cannot report requested information without records to transform data into information. The information received by the government is the input that eventually is reported out in national statistics from OSHA 200 reports. In this way, it can be said that company occupational health and safety records make a meaningful contribution to the knowledge of our well-being on a national scale.

OSHA's first requirement is that a company write a planned program for protecting employee health and safety. This is typically a manual record and is not adaptable for entry into a computer system such as this book describes. Once written, this document does not undergo frequent or substantial revision. The significance of the record is that the company has committed itself, under government mandate, to an occupational health and safety program and has described the plan. This is the first record government inspectors check when they visit company offices. It is this primary document that is most often found to be missing or inadequate, prompting citations.

REQUIRED AND OPTIONAL INDUSTRIAL HEALTH AND SAFETY RECORDS

Preplacement exam laboratory and other test results

Illness and injury records

Job summary records

Employee training

Employee name and Social Security Number
Date(s) of training
Training content

Monitoring data

Number of samples taken and duration
Concentration results
Sampling equipment used
Calibration data
Evidence of accuracy of sampling
Analytical methods and evidence of accuracy
Names and Social Security Numbers of employees monitored
Applicable respiratory protection

Ventilation and emissions records

Related control system
Measurement type
Date of measurement
Measurement results

Fig. 4.2a Required and optional records for an occupational health and safety computer system.

Although some hazardous materials and conditions carry specific requirements, the general list shown in Figures 4.2a and 4.2b covers the scope of industrial recordkeeping requirements for occupational health and safety.

Recordkeepers. A long-standing practice in business accounting is to employ a bookkeeper as a specialist for maintaining the company's accounts receivable, accounts payable, journals, and ledgers. This is a

REQUIRED AND OPTIONAL INDUSTRIAL HEALTH AND SAFETY RECORDS(cont.)

Employee exposure data

Employee name and Social Security number
Exposure calculations
Exposure observations and measurements
Physician's opinion about the employee exposure
Physician's recommended restrictions
Employee complaints and symptoms
Date(s) of exposure
Job assignment at time of exposure
Location of employee at time of exposure

Employee notifications

Monitoring and exposure measurement(s)
Require medical and diagnostic exams

Hazardous agent inventory

By location
By quantities used
By time of use

Material Safety Data Sheets

Permits

Hazardous waste generation
Transport of hazardous materials
Transport of hazardous waste

Others

Underground storage tanks and inspections
Shipping manifests for hazardous materials
Emergency response data

Fig. 4.2b Required and optional records for an occupational health and safety computer system (cont.).

recognition of the fact that it takes full-time concentration to follow cash flow into and out of company bank accounts. If all occupational health and safety records were collectively maintained, the job of recordkeeping would be equally demanding, but for valid reasons the clinic, industrial hygiene, and safety keep separate files. What should be emphasized is that it is imperative to keep the records in the best possible manner because they are mandated and a part of the company's legal responsibility, and because they need to be accurate and complete for the company's protection. The intent of the government is to protect employees and the public. The intent of the company should be to protect itself, while being a responsible employer and member of the community. Experienced recordkeepers for occupational health and safety should be provided for in a typical company budget, leaving the professionals free to concentrate on carrying out their technical responsibilities. Keeping vitally important records well is inexpensive insurance.

Archives. Preservation of the company data is archiving. OSHA requires that some records be kept for a minimum of 30 years. Forms are used to record evidence in an organized manner, and the records are then preserved by archiving. Archiving is a concept to which little attention is paid, but in an integrated occupational health and safety system it is of great importance. Just as the government amasses OSHA 200 data for evaluation of workplace illness and injury statistics on a national basis, every company should consider statistical evaluation of its collected data on a periodic basis. This can be done in-house or by contract with epidemiologists. When occupational health and safety systems are developed, there is rarely enough epidemiological input into database design. For this "down-the-road" use of the system, appropriate early suggestions should be made to promise adequate data for evaluation. Retroactive studies cannot be carried out if the input data are disparate.

Records Evaluation. From either point of view, government or company, the object of keeping records is to be able to evaluate evidence. Records are a means to an end, and if the company is required by law to have a health and safety program, it is expected to be able to evaluate the progress, success, or failure of that program. Data analysis and evaluation will describe the effectiveness of the program, reveal the existence and prevalence of problems, the rate at which program goals are being met, and the need for remedial measures. This is what the government hopes the company will accomplish on its own, but lacking company initiative, the government finds it necessary to apply oversight. Assistant Secretary of Labor for Occupational Safety and Health John A. Prendergrass has said that "Recordkeeping must be the cornerstone of any successful safety and health program" (Barbara Mabbett-Shoop, *Industrial Safety and Hygiene News*, Sept. 1987, page 65.)

Paper Files or Electronic Files. The company must decide whether records will be kept on paper in steel file cabinets or on magnetic media.

Either way, recordkeeping will be an inescapable expense. Over a period of many years, paper files will consume a lot of space and require some measure of security. It is conceivable that everything could be photocopied and the copies stored elsewhere, but the duplication of space alone would be appalling for a company of any size. By contrast, long-term storage on magnetic media is relatively easy, inexpensive, and practical. The initial effort to prepare a computer system and transfer from paper records to magnetic files is time-consuming and expensive, but it is a necessary archival procedure for the coming century which will inevitably concentrate more than ever before on accumulating data for evaluation. Manual methods will no longer suffice. Magnetic media have one disadvantage: They can degrade over time. An inexpensive solution is to copy stored files every four or five years.

On occasion, the hope is expressed that conversion from paper files to magnetic media will lead to a paperless office. This is simply not going to happen. In fact, computers generate a lot of paper from printouts of data on file and numerous reports. Printouts are necessary to check what is in the database. Things like employee lists and work assignment lists are handy desktop reference items. Quickly composed query reports on employees who have positive medical test results or exposures above action levels are useful. Obtaining query-type material from paper files would be time-consuming. Hence, the computer as an information tool is utilitarian and fast, but it will not replace paper files. The real change will be in the content of paper files, from paper forms filled with data to report forms filled with information.

Printers. When the user moves away from the computer, in all probability it will be with a piece of paper in hand to transport to another office, to supply information for a meeting, or to give to another employee. This paper has to come from a printer connected to the computer. Users and developers together select peripheral printing equipment that will suit user needs. The printer maximizes the usefulness of the computer. Just as the photocopier has exponentially increased the use of paper in recent decades, computer printers are expected to join forces with copying and faxing capabilities, setting off another exponential increase in paper consumption.

Bad Recordkeeping. While some large corporations have had recordkeeping problems of great magnitude, it is normally small businesses that have the worst recordkeeping reputation. This is partly for lack of personnel and equipment and partly for lack of genuine concern for the role of occupational health and safety records. Few small companies can afford automation beyond a microcomputer and inexpensive software, and many contract out medical and industrial hygiene services, expecting the contractors to keep records while providing professional services.

Worse yet, too many companies, large and small, play games with their records and falsify injury and illness data. A report in *Industrial Safety and Hygiene News* of September 1987, page 79, says that 69% of some readers polled felt that falsification was done to help "control insurance and compensation costs," and 53% said it was done to "avoid OSHA inspection by having an accident rate lower than the national average". No percentage was given for the number of companies involved, only the reasons for falsification. Much of the poll appeared directed at the OSHA 200 and 201 report forms. Part of the poll indicated willful reporting violations because companies deliberately kept bad records.

Nothing will justify deliberately bad recordkeeping. All companies do not have the ability to pour money into occupational health and safety, but every company can exhibit a willingness to deal honestly with the health of its employees. There are no quick fixes, but recordkeeping can do a lot to reduce hazardous conditions. Those companies that have kept good records and used them to measure their workplace health and fitness have shown repeatedly the value of both programs and procedures for evaluation. The will to protect and improve begins with employee involvement in an occupational health and safety program coupled with the determination to measure improvement through good recordkeeping.

For too many years, the occupational health and safety field has suffered from a lack of good recordkeeping tools. There is a genuine need for integrated computerized systems for microcomputers, minicomputers, and mainframes at an affordable price for companies of all sizes. When recordkeeping in this field improves, services provided for employees will improve, and we will come closer to achieving epidemiological evidence of where the greatest exposure hazards exist in industry and how to prevent them.

4.3 HELPING THE DEVELOPER WITH SYSTEM DESIGN

Physical Arrangements. The analyst will need a tour of the physical facilities. Where will the equipment be located in the departments, where are the electrical outlets and communications lines, and who will be using the equipment and when? This tour is the analyst's opportunity to view what exists and to envision what can be done prior to recommending hardware selection, system configuration, and security requirements.

The analyst will also make a request to talk with each of the users, including personnel who may be collecting the data but not actually using the computer.

Collecting Forms. Only users will be able to judge what data should be captured. This is not the developer's decision. It is the developer's obligation to advise users about ways to improve data capture, processing, and retrieval. Once actual design of the system begins, users will be asked to collect all paper forms used in recordkeeping, reviewing each one, item by item, to help the developer understand the flow of the data, the contribution of each data element to the total information picture, and perhaps some of the requirements for using the data (like government mandate, legal protection, company policy, or use in evaluations). The developer will need time to assimilate the massive data demands of an occupational health and safety system, which is probably a new experience.

The Questionnaire Problem. Immediate concerns of the developer will arise when the user presents a questionnaire-type form. The questionnaire that comes to attention first is the medical history form. How is the record to be input, stored, and later updated? This is usually a long record which can be broken down into smaller sections, such as family history, exposure history (occupational and nonoccupational), life-style, and medical history. The next problem is how to enter answers to the questions. Does an employee or a nurse make the entries? Are entries made on a paper form or directly into the computer?

What kinds of answers are expected? Will everything be "Yes-No" replies, and what happens if an answer is not known? Perhaps a third response, "Uncertain," is appropriate; options would be single characters like "Y/N/U." Frequently, categorical questions may be followed by "Other?" This requires some short text entry and may need 15 or 20 characters for a response. Some questions will need short text responses to explain a positive answer. For instance, a specific illness might carry a short line for entry of the year or the number of years ago. User and developer will spend considerable time resolving the many small issues that surround questionnaires. Figure 4.3a illustrates a medical history page.

The person who fills out the form also determines the language of the form. For employee use, medications may be described in layman's language (e.g., blood pressure pills), but if a nurse fills out the form, a medical term will be used (e.g., antihypertensive). While these concerns may be commonplace to medical professionals who are accustomed to writing questionnaires, the developer must explore every aspect of tailoring the medical history form to a new medium.

The most difficult question will be how to update the record without losing any earlier responses. For instance, on the preplacement and periodic medical histories, a question about ringing in the ears may have

```
MODULE: Medical                                        PROGRAM: Exam
                        MEDICAL HISTORY

EMPLOYEE ID:                                   NAME:
      DATE:
-------------------------------------------------------------------
Have you ever had:  (0 = No, 1 = Yes, 2 = Uncertain)

_ Allergies                _ Glaucoma           _ Rheumatic fever
_ Anemia                   _ Heart attack       _ Rheumatism
_ Arthritis                _ Heart disease      _ Seizures
_ Asthma                   _ Hepatitis          _ Shortness of breath
_ Back problems            _ Hernia             _ Stomach ulcer
_ Bleeding disease         _ High blood press.  _ Sinus trouble
_ Cancer                   _ High blood sugar   _ Stroke
_ Chest pain               _ Juandice           _ Thyroid disease
_ Convulsions              _ Kidney disease     _ Tuberculosis
_ Diabetes                 _ Liver disease      _ Varicose veins
_ Fainting/dizzy spells    _ Nervous breakdown  _ Venereal disease
_ Gall bladder trouble     _ Polio              _ Xray treatment
```

Fig. 4.3a An illustration of a medical questionnaire screen form.

had a negative response, and five years later it has a positive response. Some users prefer to record only positive responses; others prefer to record all responses. How will the earlier data be preserved, and how will the change be recorded?

Another interesting concern of a similar nature arises with the physical exam. A typical issue is how to add the doctor's comments to conditions noted during the exam. It will be up to the developer to devise

```
MODULE: Safety                                          PROGRAM: Audit
                          PLANT SAFETY AUDIT

AUDIT NO: 5                                        DATE: 04/18/89
LOCATION: ABC Plant                             PURPOSE: AA Annual Audit
-----------------------------------------------------------------------
Training: (0=Not applicable, 1=Unacceptable, 2=Needs improvement,
                     3=Acceptable, 4=Excellent)
                                                                  R  C
Accident prevention training for: all employees. . . . . . . .   _  _
                                  all management . . . . . . .   _  _
Supervisory training for: Accident investigation . . . . . . .   _  _
                          Hazard identification. . . . . . . .   _  _
                          Leadership . . . . . . . . . . . . .   _  _
                          PPE use and care . . . . . . . . . .   _  _
                          Recordkeeping. . . . . . . . . . . .   _  _
                          Safe work practices. . . . . . . . .   _  _
                          Safety committee work. . . . . . . .   _  _
                          Safety and health requirements . . .   _  _

Rating (this page):       Maximum 40                             ____
```

Fig. 4.3b An illustration of a safety questionnaire screen form.

a method for capturing narrative comments along with the "normal," "abnormal," or "not examined" responses common to a physical exam.

In safety management, a plant audit can be a questionnaire similar to the medical history. Questionnaires present special concerns in regard to format, screen presentation, response entry, storage, and retrieval. On the computer, it is advisable to fill the screen rather than deal laboriously with one question at a time. Opinions about storage will differ, as suggested above. Retrieval questions are similar to those of storage. Should only positive responses be retrieved, or should all questions be brought back to the screen? Safety audits may carry ratings instead of positive or negative answers see Figure 4.3b.

In using already written safety audit and medical history questionnaires in a copyrighted software system, make certain that copying them into the system does not abridge someone's copyright.

An issue with regard to questionnaires in computerized systems is how to cope with changes to the form over time by adding and deleting questions. One possible solution is to have a dictionary table of questions that are coded, numbered, and dated. When the questionnaire is called up, a table search brings the questions onto the screen by code and number, presenting a screen questionnaire form whose lookup origin is transparent to the user. The developer has to store the record under the employee's name, identifying the questions that were used.

Calculations. The developer will need all help possible from users on some of the calculations in occupational health and safety. Being unaccustomed to concepts like time-weighted averages (TWAs), corrections for presbycusis, significant threshold shifts (STSs), and conversions from parts per million (PPM) to milligrams per cubic meter (MG/M3), the developer will require patient instruction in the variety of required arithmetic procedures. (Uppercase letters, superscripts, and subscripts in units are discussed in Section 4.6.) The user should explain the concept and the place for the procedure, but should also write out the procedure for the analyst and programmers to follow, step by step. Simply handing the developer a formula, unless it is an elementary one, can lead to misinterpretation. A good programmer can condense many arithmetic operations into a single long computer formula, but the sequence of operations must first be clearly understood. The user can best ensure computational accuracy by spelling out the sequential operations, adding instructions where necessary to look things up in tables for further help. Lookups would be used in prebycusis corrections, predicted values in pulmonary function tests, and the formulas to plug in for conversions of units. The analyst and programmers have to translate normal arithmetic notation into a language and format that the computer understands.

Pulmonary function test (PFT) calculations are an excellent example. All of the PFT calculations for an exam will not be given here; only certain ones are needed to illustrate the point just made. Calculation of predicted normal values, shown below, presumes an employee record containing the sex, race, and date of birth, a standard routine for computing the age wherever it is needed in the system, height as given in the PFT record, and a lookup table of BTPS values for machine temperature correction. The formulas are from the American Medical Association's *Guide to Evaluation of Permanent Impairment,* April 1984). Convert height from inches to centimeters

1. Height (cm) = Height (inches) multiplied by 2.54
   Round up to next integer
2. After entry of raw value, only if Race = Black
   Corrected FVC = FVC (raw) divided by 0.9
   Round to two decimal places
   Hereafter, use Corrected FVC as FVC (raw)
3. Look up conversion factor for given PFT machine temperature in BTPS table
   Corrected temperature = FVC multiplied by conversion factor
   Round to two decimal places
4. If Sex = Male
   Predicted Normal FVC = (0.0600 multiplied by Height (cm)) - (0.0214 multiplied by Age) - 4.650
   Round to two decimal places
   If Sex = Female
   Predicted Normal FVC = (0.0491 multiplied by Height (cm)) - (0.0216 multiplied by Age) - 3.590

Not Everything Needs to be Computerized. Ideally, a computer system should be as handy in the daily office routine as pencil and paper, but computers fall a little short of the ideal. With several applications packages loaded into the machine, like a word processor, a spreadsheet, and an occupational health and safety system, a lot can be accomplished. Letters, memos, and safety committee notes can be done on the word processor, computations can be done on the spreadsheet, and occupational health and safety can process most technical data.

However, the occupational health and safety system should not become a catchall. Discretion should be used to keep this from being a storage shed. For example, codes are important to the system (for reasons that will be discussed later in this chapter), and use of standard codes like ICD-9 is encouraged. Tapes of these codes are available for purchase, but most of the codes will never be used. It would be more practical to keep an ICD-9 manual handy, initially entering as many codes as seem useful. From time to time, more codes can be added to the code tables as needed.

MSDSs number in the thousands for big companies and require a lot of entry time. Commercial software is available for MSDSs, but smaller companies probably need only a fraction of them. It is worth the time and money to do a large one-time entry of those that are usable, adding only as required rather than keeping a large database on hand, many of which are useless.

A problem presented by MSDSs is how to revise and maintain them. The user should discuss this thoroughly with the systems analyst so that the system design will resolve the problem. It is senseless to keep revision after revision on hand. Print out an MSDS before making a revision, date it, and store it in a paper file rather than keep every revision of every safety data sheet in the computer. Break up the MSDS into sections for convenient entry and revise only in the necessary sections. Not all MSDSs are 10 or 15 pages long, but design the system so that, short or long, every one is easily handled.

If it is desirable to have very detailed training course contents, use the computer record for a condensed description of the course and keep a lengthy syllabus in the training materials file drawer.

Decision Making. An occupational health and safety applications system processes data for informed decision making. It is not an artificial intelligence (AI) type of system for decision support. An expert system is also built around a database, and it is used for making educated programmed decisions. Information from the occupational health and safety system is used by humans to make their own decisions; in this system, data are organized so that they can be retrieved in an efficient, time-saving, and cost-effective manner. The expert system is programmed to make sequential tests of the data to meet programmed standards; at the end of this process, the computer arrives at a conclusion. The tasks are perfunctory and repetitive, each test result being based on the conditions as presented by the programming. This is a decision-support system.

Decision-support systems are still in their infancy, and not all are tied to AI. Some commercially available products integrate into a single package a DBMS, an electronic spreadsheet, a word processor, and graphics. In time, they may become useful in occupational health and safety for evaluating hazardous conditions, exposure records, and the effectiveness of control systems. Through interfacing, the occupational health and safety database could contribute data for statistical analysis, but incorporation of decision support is only a future possibility.

4.4 FORMS BECOME SCREENS

Users Design Records. To say that users design records could be misleading, but the fact is that it is through the eyes of the users that the analyst is able to envision the requirements. What follows is a system design and a screen design that matches what the user has outlined in detail. An occupational health and safety system arises from the data keeping of medical and safety specialists and industrial hygienists. Users should adapt the system to their work practices, technical language, and scientific data. There are so many special requirements that ordinary software applications are not adaptable.

Analysts always launch system design by collecting copies of every paper form employed by the users. Paper forms have been designed by users to organize facts that must be preserved for future reference. A form is a document that stores data. It becomes a record in the paper system in which facts are recorded in an orderly manner, stored in a file cabinet, and retrieved as users need information.

Information. Organized records avoid random and unmethodical perusal of the facts, an indispensable ingredient of all sciences, pure or applied. Perusal leads to an analysis that provides information. Information then becomes a synthesis of facts in the file that enlightens, instructs, reports, notifies, or intimates and that may be used to inform, prompt, or confirm in either a negative or a positive way.

School children learn about the scientific method, in which a condition is identified, observed, and recorded in detail. The observations are analyzed, an explanation for the condition evolves from the analysis, and conclusions are drawn about the reasons for the condition. This is precisely how occupational health and safety works, although we seldom see it from this elementary perpective. In designing the occupational health and safety system, a lot of mistakes can be avoided by generating a database from its embryonic stages.

The condition to be addressed is the preservation of health and safety in the workplace and the environment. The observations are medical and industrial hygiene test measurements and safety incident investigations, which are later analyzed and reported. The conclusions that follow are a synthesis of all the information that confirms the absence or presence of a condition. A pivotal point in the process is the recording of observations in order to create a body of facts.

Facts. In any scientific investigation, the facts are recorded when observed; otherwise, accuracy diminishes quickly over time. Unless recording is automated, most of these observations are made manually on paper forms which accumulate into an uncontrolled collection of facts. Ideally, occupational health and safety would benefit by entering all measurements directly into the computer, but that has been more hope than reality up to the present. Some laboratory instruments can be interfaced to a computer, but the greater part of recording is done by hand and entered into the computer later, inasmuch as computers are not generally taken to the field. However, the computer has come of age, and it is time to interface recording machines to an occupational health and safety system.

Users design the forms for recording the facts (data). Care in design will be rewarded with reliable data for later analysis. The form has to be designed with several features in mind: order of recording, type of data to be recorded, understandability, usefulness, and the manner of drawing conclusions.

Suggestions for Designing a Form. Three primary requirements for designing a form are to determine:

- Who will fill out the form.
- The conditions under which the facts will be recorded.
- Which facts must be preserved.

Several points about who fills out the form are worth consideration. Forms that use technical language should be filled out by professionals who are familiar with the data. However, forms that need to be filled out by people with a variety of backgrounds must be brief and written simply and clearly. Dyslexics read with difficulty, bilingual employees need simple English (and perhaps their original language), and poor spellers should not be expected to write essays in response to a question. While medical personnel who are familiar with their own jargon fill out their forms, and industrial hygienists understand the manner in which they enter data into their forms, safety professionals often do not fill out accident investigation forms. Supervisors and foremen provide this information and can do the best job with an uncomplicated form.

Conditions in the field may not be favorable for completing a form in its entirety; this is no problem. Where conditions prohibit full recording on the scene, complete the records as soon as possible afterward. Back at the desk, narrative observations can be added from data recorded at the time of field observations. However, one situation where on-the-scene narrative is important is an incident investigation. For instance, descriptions by witnesses should be recorded in full. If narrative writing is a problem, use a tape recorder and later transcribe important facts into the record.

Forms design can easily get out of hand; it is not as easy as it looks to produce a truly good form. A long form may be improved by breaking it down into two or more shorter forms—for example, an incident investigation form that includes not only the details of the incident, but also safety committee recommendations and corrective actions. There are problems with making several forms out of one, chief among which is how to keep all the related forms physically tied together. This can be done by numbering them all the same way or physically filing them all together. System design will keep them together.

Long forms are not likely to be filled out all at once; they may require several attempts at completion. They tend to be tedious because they require so much detail. Narrative answers require thought. A question like "Was there a violation? Explain." will require a lot of text. The first fact is simply "Yes" or "No", but the explanation of that fact may be quite involved. This creates the problem, mentioned in Chapter 3, of using text in the database and burdening computer storage space. Too many questions followed by "Explain" or "Specify" produce an overloaded database.

Short forms can be dangerous for lack of data or too abbreviated data. One supervisor was describing his company's accident investigation report. "Shucks, I can fill that thing out in fifteen minutes. It's gonna do nobody any good 'cause there's nothin' to it," he lamented. This was a metal processing plant. If the pot line supervisor was contemptuous of the company's records, he either did not understand the investigation process or the record was truly poor. If he did not understand the necessity for the data, training could have been at fault, or he may have been the wrong person to be filling out an important document like an incident investigation. Good records are not a happenstance; they are painstakingly created.

Developing a Form. The place to begin is with the user's staff, who will supply the bulk of the required data. Next, find out what management and related departments think would be helpful to them. Some good suggestions can be made, especially relating to the business side of the company. Large companies require forms control, necessitating some red tape and effort before adopting a new form. A third resource is professional organizations, where colleagues are usually full of ideas about their own experience and can supply a list of dos and don'ts. They may also offer copies of their own forms.

A normal by-product of designing a new computer system is the opportunity to redesign forms and work practices that should have been revised before. The design process should stimulate users to think anew about what they have been doing and how improvements can be made. As disruptive as this may be, most users welcome the opportunity to improve their records.

When the form has been drafted, test it on user personnel, using real data, especially data at the extremes. A surprising number of errors can be caught through this exercise even though everyone has had input to the form design. Manually attempt each of the following for the problems that might arise:

- Add a whole record.
- Add new data to an existing record.
- Change existing data in a record.
- Delete data from an existing record.

Training. Training people to fill out forms is almost as important as designing the form. The pot line supervisor who had no respect for his company's accident investigation form could well have misunderstood the importance of the information he was asked to give. Every user of a form should be conversant with each item on the form and understand how the company intends to use the data, as well as what to put in the item.

Screens. Users design forms, and forms in the hands of the developer become screens. The user interface is the computer version of a paper form and should replicate the paper form as closely as possible. When paper forms are returned from the field, the data should be entered into the computer system shortly thereafter. Users must decide whether the paper form will be retained on file temporarily or permanently, in addition to the computer record.

Screen forms have a few disadvantages over paper forms. One is that the preparer of the form cannot affix a signature or certification stamp. This problem can be overcome by entering the person's name, certification ID number, and date into the computer, while the paper form with the signature can be kept on file. The real disadvantage of the screen form is that input time is doubled when a paper form is copied into the computer. The future solution will be to enter as much data as possible directly into the computer.

The logical advantage of records is not just that they hold data, but that they allow users to retrieve and process data for output in the fastest and easiest way. The computer cannot do anything the user cannot do for himself, but it retrieves and processes faster and stores in much less space than paper files. An organized database makes it easier to find data upon request.

An Illustration of a Screen Form. Figure 4.4 illustrates some desirable features in a screen form and some features that require the special attention of the developer. The top line is the status line that defines the user's present location in the system by module and program. In the same block is the title of this screen, all of which is given by the system in response to the user's menu selection. The area for data entry is grouped into blocks of related data, each group separated by a line of dashes. Prompts are all in boldface type, initial caps, followed by a colon and a space. Data entered by the user are entirely in uppercase letters and have a different screen intensity than the prompts. Dates have delimiters entered by the system.

When 101 is entered for the Course ID, the system does a lookup to the Course ID table and brings back the description of the course. The frequency of the course offering is to be given in weeks, a unit that automatically comes up as a default value. This acts as a prompt aid for the user, but it can be overridden if months or years are preferred. Default values are available at the user's discretion. They need to be programmed in but are seldom a problem. If no default is given in the illustration, some indicated period of time should be shown in the prompt in order to keep the records uniform.

Certification is a coded entry. The list of codes for certification could be:

```
Module: SAFETY                                     Program: TRAINING
                         TRAINING COURSES

Course ID: 101               Description: HANDLING SOLVENTS
Frequency:  26 WEEKS       Certification: 01 NONE REQUIRED
--------------------------------------------------------------------
Content: T       Number of Hours:    3        Max. Enrollment: 10

             Materials Required: T        Other Requirements: T
--------------------------------------------------------------------
             Budget
                     Materials:     45
                    Instructor:     60
                         Other:     10

                         Total:    105
--------------------------------------------------------------------
Requested By: B. CALLAHAN              Approved By: C. KROGER
        Date: 02/07/89                        Date: 02/10/89
```

Fig. 4.4 An illustration of a screen form.

01 None required
02 First Aid
03 CPR, etc.

When the cursor reached this field, the user entered 01, and the system looked up the code and brought back the code description NONE REQUIRED.

In fields carrying a "T," text has been entered for that prompt. The text narrative is filed separately after having been entered in a window. Entry and filing mechanisms must be designed and programmed by the designer and programmers.

Blocks often benefit by having a name for the grouped data, but many blocks in the occupational health and safety system contain such a miscellany of data that this practice need not be a hard and fast rule. The budget block in the illustration lends itself to a block name. Since this budget is intended only as a reference record, the figures are rounded off to the nearest dollar. The field descriptions would be given in whole number values. Once entries are made in all the fields, the system automatically does the addition and prints the total. "Total" is a computed variable.

Paper forms are normally approved and signed. Lacking the ability to incorporate signatures into a computerized form, enter a name, ID, or code of some sort with a date.

4.5 WORK PRACTICES AND SYSTEM DESIGN

The Program Concept. Activities in occupational health and safety are commonly referred to as *programs;* even the health and safety effort is a company program. There are subprograms like monitoring program, training program, fitness program, and hearing conservation program. Each of these represents a set of tasks for which there are unique records. For illustration purposes only, the occupational health and safety system has been divided in this book first into modules, then into programs, and lastly into screens, which are replicas of paper forms. The nomenclature used should not inhibit the reader from using whatever terminology is more typical of his or her company's language.

In Section 3.4, the analyst began to design the system by grouping user tasks into programs. As the user assists in this design phase, there is ample opportunity to reexamine work practices for potential improvement and to add a wish list of new activities. Even if it is not feasible to initiate a new program at present, add it to the design to see how it fits.

After all, this is only a paper drawing. The advantage of brainstorming in this early design phase is that it provides the growth potential that the system will require if it is to remain useful in future years. In the occupational health and safety system, new programs are fast becoming popular, especially in the health area, examples of which are fitness, weight loss, and employee assistance. Safety management is interested in back programs and off-the-job training programs. Ultimately, expert systems and specialized software will be added to the basic health and safety system. Like building a house, it is more economical to make the attic high enough in the planning stages than to change the roof when the time comes to add more bedroom space.

Basic Documentation. It is best if users present written outlines of their programs. As the analyst receives written guidelines from the users, information about the system will accumulate, establishing some basic documentation materials to accompany design descriptions. This will be of future value.

Procedures manuals are of great value to analysts during systems design. They seem to be more common in clinics than elsewhere in safety and health. So much about work practices has to be described for analysts that a valuable by-product could be the development of a procedures manual, if none already exists, or the updating of those that do exist.

Scheduling. Of all the professional activities in occupational health and safety, the most intractable problem concerns the development of a scheduling program. Every effort thus far has dealt with the issue in a limited way. While commercial developers advertise scheduling programs, they have never offered a true scheduling system. At one time, a prototype system for a microcomputer was implemented, but it required more time for debugging than was feasible for commercial development. The conclusion is that doing scheduling by computer is possible, but at present we lack the expertise to do it cost-effectively.

Even in the offices of physicians and dentists, receptionists scheduling appointments manually in a big book, while their other tasks are computerized. The scheduler works in pencil because there are so many possibilities for change: Patients cancel or switch appointment times; doctors and nurses take sick leave, are called for emergencies, or are not available as scheduled for various reasons; emergency patients need to be squeezed into already crowded schedules for immediate attention.

Computers are not as yielding as the appointment book. They are designed to receive and store data. Temporary data entry is not a normal function, but in the appointment book, every entry is temporary until the patient has completed the appointment, whereupon data from the visit can be stored in the computer.

While it seems logical that a framework of 5-, 10-, or 15-minute intervals for each day could be blocked out and names entered, the programming is not that simple. When a change is needed, you cannot simply delete it from one block and add it to another because the new appointment may be many days away. In the appointment book, you can view a week at a time. On the screen, you may view only a day or two at a time. The programming challenge is to call up a new available time on a new available date that is not on the screen. In other words, the computer must simultaneously keep track of all available open time from the lists, delete scheduled time and add it back to open time, and remove the open time. The two key concerns are limited screen space for a very large page, and maneuvering data into and out of time slots quickly and efficiently.

Schedules can be entered into the computer each day after completion in the appointment book, but complete elimination of the book is not desirable. In the occupational health and safety system, scheduling is a typical user task for the clinic, industrial hygiene, and safety. The clinic schedules medical exams, the hygienist schedules surveys, and the safety department schedules employees for training. In safety and industrial hygiene, even equipment may be scheduled. Each of these departments has it own scheduling functions. If the clinic has more than one examiner, the schedule must show not only the employees but their assignment to an examiner. Examination rooms may be another parameter for consideration. Industrial hygiene is concerned with scheduling personnel and equipment to specific locations. The safety department schedules employees into courses, courses into time slots, and time slots into available training locations. When a developer resolves the complex programming problems of scheduling, tailoring features for each user still remain. Scheduling is a desirable feature of an occupational health and safety system that may continue to be done in appointment books for lack of sufficient computer resources. The usual system solution is to keep an appointment book and enter the daily schedule after completion. This is not automating the scheduling task; it is only using the system as a recordkeeper after the initial task is completed manually.

Selecting Employees. The computer system is adaptable to selecting individuals and locations for medical exams, monitoring, and training. This can be done by query, employing user-specified conditions, or by programming the task directly into the system. Since automation of occupational health and safety is the objective, programming the selection process into the system should be the appropriate system design.

This is the place for prioritization, as suggested in Section 4.1 and discussed further in Section 4.8. Prioritize employees and locations for surveillance according to their potential for exposure within a job class and the toxicity of an agent. The objective is to assess risks in priority order so that resources can be allocated wisely. This encourages systematic medical exams and industrial hygiene sampling strategies that will produce reliable exposure profiles. The process works best when coded inventories of jobs and agents are available.

Group employees by similar exposures and agent toxicity. Homogeneous exposure groups (HEGs) and "zoning" are not universally used, unfortunately, but implementation of a computer system is a good time to begin. Once organized, the groups can be ranked and each employee assigned to a job with an exposure risk rank. The computer can be given the task of determining when clinic exams and sampling surveys are to be made, and notices can be sent. A selection system based on jobs and agents supports periodic medical exams, industrial hygiene monitoring, and safety training so effectively that it is worth the cost of development.

Ticklers. Another task common to all three disciplines is a reminder ("tickler") file. To ensure that actions are carried out as planned, follow-ups are important to everyone (Figure 4.5a). In the clinic, desk calendars are used when employees are referred to outside medical services, returned to work, and in need of medical checkups. Tickler files of 3 × 5-inch cards provide a reminder of when employees need to be brought in for tests. Calendars in industrial hygiene and safety carry dozens of reminders, and tickler files remind about calibrations and training courses. Whether the reminders are a day, a week, or a year away is immaterial; the computer can improve upon this practice.

A single tickler file for all departments would be impractical, and each department should have its own. Sharing a database does not require that every file be shared. Where work practices are specific to a department, it is best to limit access to department files.

Order of Data Collection. The analyst may want to observe personnel to obtain information on work practices. This is especially important if on-line entry is to take place. All screens should be arranged in the order in which data are taken, a point that was made earlier with regard to forms design. As an example, when someone enters the clinic and undergoes routine procedures like taking the pulse rate, blood pressure, and a weight check, what is the order in which these things are done? Figure 4.5b is an

```
MODULE: General                                          PROGRAM: Tickler
                              TICKLER FILE

TYPE: 01 Med. Follow Up              DESCRIPTION: J. Jones Referral
PRIORITY: 3 Routine               PRESENT STATUS: 1 Incomplete
-------------------------------------------------------------------
DATE REQUESTED: 03/01/1989             REQUESTOR: Dr. Smith
-------------------------------------------------------------------
    TARGET DATES:  Begin                  Complete  04/01/1989

    ACTUAL DATES:  Begin                  Complete
-------------------------------------------------------------------
PERSON RESPONSIBLE: B. Anderson                    COMMENTS:
```

Fig. 4.5a An illustration of a tickler file screen form.

example of a vitals screen, reflecting one sequence of these common procedures. Although this screen is a separate record, it is also possible to have a clinic visit screen that includes these data. The screen should reflect the order in which the nurse gathers the data, especially if on-line data entry is used during the examination.

```
Module: MEDICAL                                        Program: CLINIC
                                VITALS

EMPLOYEE ID: 12345           EMPLOYEE NAME: David Stephens
AGE: 34      SEX: M
VISIT NO: 28                 REASON: 6 BLOOD PRESSURE MONITORING
------------------------------------------------------------------
EXAM DATE: 03/08/1989                              EXAM TIME: 10:08
------------------------------------------------------------------
TEMPERATURE (F): 98.6                              HEIGHT(ins):  73

PULSE: 86                                          WEIGHT(lbs): 171

BLOOD PRESSURE:
      Position:               Arm:          Systolic / Diastolic
            S                   L              95          135
            L                   R              90          135
            _                   _              ___         ___
------------------------------------------------------------------
EXAM COMMENTS:                              EXAMINER: Crosby
```

Fig. 4.5b An illustration of sequencing data for a screen form that matches normal work practices.

4.6 INPUT SPECIFICATION

Input Specification. Input specification is data specification that helps the user describe for the analyst how the data should look at entry. We have already discussed the data dictionary, which may be putting the cart before the horse, since the analyst will use input specifications to design the database and construct the data dictionary. The user provides a variety of information about input and output data. (Output data will be discussed in Chapter 5 in connection with queries and reports.) As users work with the analyst to describe data in paper forms, lists of files and the data they contain emerge, and the analyst and programmers will give each element a system name. The user will determine the longest entry necessary for the field. Fields containing employee names and addresses vary to such a degree that an optimal length is arbitrarily assigned rather than attempting to guess what the longest name will be.

Employee identification is always an issue. Usually a name length of 30 to 35 characters will suffice, but questions invariably arise about how to capture them. Should one, two, or three fields be used, should middle names or only middle initials be used, and should nicknames accompany names? In some plants, records are filed by name in clinical files and by social security number in industrial hygiene and safety files.

Field Types. Field types were described earlier as alphanumeric, integer, and real. Common designations for these are A for alphanumeric, I for integer number, and F for a floating point decimal, or real number. The floating point decimal must be described with care. The length of the field includes the decimal point. A common method is to indicate the field length and the position of the decimal points. For example, F6.2 says that this is a floating point decimal number of five characters, plus the decimal point. The decimal point is followed by two decimal places. Thus, F6.2 might represent 187.65.

Alphanumeric entries are normally left justified, but numbers are usually right justified; this is generally done by computer systems automatically. Consider the example of an F6.2 number. The entry 187.65 fills the field, but if the entry were 7.65 or 87.65, left justification would misplace the decimal point. Right justification always locates the decimal point two places to the left of the right end of the field.

Real Number Data. Files containing laboratory results for medical and sampling data commonly contain real numbers. A floating point may vary among different types of tests. Data entry can be facilitated if screen formats are organized for either real or integer numbers. For instance, a cholesterol reading of 175 mg/dl would be entered as 175.000, and a creatinine level would be entered as 000.600. The field length would be F7.3 and would carry any blood chemistry test results.

Industrial hygienists frequently have very small trace concentrations that exceed accurate measurement limits. They commonly use a Boolean character to show something as less than a lower limit. A field of a single Boolean character should be included in the file representing $<$, $>$, $-$, $\leq$, $\geq$. Some Boolean notation is unusable in computers and printers. It is easier to read letters for operators such as LT (less than), GT (greater than), ET (equal to), LE (less than or equal to), GE (greater than or equal to), and NE (not equal to).

Units. One of the vexing small problems to solve for laboratory test files of all types is how to write units. This is a matter of user preference. Showing superscripts and subscripts on the screen is not possible with most software. It is easier to decide on some conventions that everyone can follow and put them in a user's reference manual. A few examples of the way units are entered are shown, explaining why user conventions need to be adopted:

Milligrams per cubic meter	MG/CM	Mg/M3	Mg/CuM
Micrograms per 100 milliliters	MCG/100ML	UG/100ml	ug/100ML
Milliunits per milliliter	milliunits/ml	I.U.	
Cubic micra	CMicra	CuU	3U
Milliequivalents per 24 hours	MEq/24hr	MilliE/24hr	mEq/24hr
Cubic millimeter	CMM	CuMm	CMm
Parts per million	PPM	ppm	
Fibers per cubic centimeter	F/CM3	Fibers/cm3	FibersCuCm

It is better to use uppercase letters for liter because the lower case "ell" can be mistaken for "one". While units should be shortened by abbreviation, some will still present problems, such as hematologic cell counts in millions per cubic millimeter (millions/mm3). Unfortunately, universally accepted ways to abbreviate most units do not exist. In an integrated occupational health and safety system, conventions ought to be agreed upon by all three disciplines.

Expressing units like those shown above in uppercase letters may be a wise decision. Among the direction to the system that the developer provides, will be a specification for entry to be recorded entirely in uppercase or a combination of both upper- and lowercase. It helps the data entry operator not to have to shift between lower- and uppercase by speeding up entry and eliminating another source of error.

Superscripts and subscripts, such as those used in chemical formulas and units, are normally provided by function key programming in word processing. To make the occupational health and safety system less complex, it is advisable to eliminate these refinements and to adopt working conventions for expressing chemical formulas and units on a single line.

Relating Data Elements to the Database. Each item needs to be related to the rest of the database. For instance, will this element be needed to locate a record? Will two or more elements constitute a unique search key? Help the analyst build a list of file keys. Keys are the unique means of access to records.

Is this a required data item without which the record would be incomplete, or is the field optional? If the screen is programmed to require certain fields, the user cannot save the file without this data item. Keep fields optional if data may need to be added later. For instance, for weather conditions, barometric pressure and temperature may be required because they must be used in sampling calculations, but wind direction and wind speed are not vital and can be optional entries.

Dates. In this period, when we are beginning to think in terms of the twenty-first century, users should give some thought to year dates. Once past 1999, the next number will be 2000. Dropping off the first two digits may cause problems. Theoretically, the occupational health and safety system is a permanent historical record and could last for 100 years or more. We are so in the habit of using only the last two digits of a year that we may need to think beyond the short term. At least 30 years of recordkeeping has been mandated, so thinking ahead is not impractical. Perhaps it is time to use all four digits for the year.

In Canada and other countries dates are written by year/day/month; in the United States we use both month/day/year and year/month/day. The analyst can state how the computer sorts an index of ascending dates and if 00 will pose problems.

Material Safety Data Sheets (MSDSs). Safety data sheets present many different data specifications. Some of the physical properties that carry numeric data cannot be entered in numeric fields because of attached conditions. For instance, solubility in a specified liquid may read "510 ppm @ 25°C." It is best to make solubility fields alphanumeric and 15 or 20 characters long in order to include the temperature-dependent feature.

Some entries will use Fahrenheit and some will use Celsius. All computers and printers do not allow for the degree mark without extended character sets. Therefore, omit the degree mark and enter "F" or "C." Flash points will need to qualify numeric measurements as open or closed cup (OC or CC), prompting a choice between one numeric field and one alphanumeric field, or one alphanumeric field. Using a minus sign in a numeric field is permissible so that boiling, melting, freezing, and flash points can be numeric fields without an additional field for the minus sign. Where certain units are constant, such as "mmHg" or "atm" for vapor pressure, use a default value or have the screen permanently marked for the unit. In MSDS entries where units always appear with a field, like "ppm" or "%", units can be written on the screen and do not have to be stored values. Figure 4.6a shows fire and explosion data.

Many of the entry requirements for MSDSs can be reduced to codes. One example is emergency measures. For first aid, there are specific emergency measures that can be coded in order to standardize the instructions and reduce data entry typing. Typical examples are: "wash promptly upon contamination", "use eyewash, soap flush skin, and obtain immediate medical attention". Other repetitious entries that can be coded are routes of entry and effects, protective measures, and PPE.

```
MODULE: Safety Data                                          PROGRAM: MSDS
                    MATERIAL SAFETY DATA SHEET
                     Fire and Explosion Data

AGENT ID: 0234 XYZ Paint                   MANUFACTURER: Chemco, Inc.
  CAS NO: 11-23-4                            MSDS DATE: 03/10/1988
---------------------------------------------------------------------
 FLASH POINT: - 106 F ___ C                    NFPA CLASS: IA Liquid
 TEST METHOD: CC                                      LEL:  2.3 %
AUTOIGNITION:    175 F ___ C                          UEL:  9.5 %

EXTINGUISHING MEDIA: FE 101 Waterfog
                     FE 104 Foam
                     FE 110 CO2

FIRE FIGHTING PROCEDURES: For large fires use water spray
                          Use respiratory protection
                          Apply fog downwind

UNUSUAL HAZARDS: FH 222 Fire causes toxic fumes
                 FH 215 Irritant
```

Fig. 4.6a An illustration of data specifications for an MSDS screen form.

```
MODULE: Safety Data                                          PROGRAM: MSDS
                         MATERIAL SAFETY DATA SHEET
                            Toxicity Information

AGENT ID: 0234 XYZ Paint                    MANUFACTURER: Chemco, Inc.
  CAS NO: 11-23-4                             MSDS DATE: 03/10/1988
-----------------------------------------------------------------------
EFFECTS: TE 123 Irritant
         TE 221 Suspected liver injury

REFERENCES: TR 567  Smith,C.G.    XYZ, rabbit open skin test
            TR 453  Brown,R.B.    XYZ, mouse inhalation test
            TR 345  Short,A.F.    XYZ, human, liver injury
-----------------------------------------------------------------------
COMMENTS:
```

```
MODULE: Safety Data                                        PROGRAM: Tables
                              MSDS REFERENCES

REFERENCE NO: TR 567   DESCRIPTION: XYZ, rabbit open skin test
      AUTHOR: Smith, C.G.
       TITLE: XYZ Paint exposure in rabbit skin test
 PUBLICATION: Amer. Coatings Jour. 45(6):444-449, 1986
```

Fig. 4.6b A suggested method for citing references in an MSDS screen form.

Citations for toxicity references require a lot of entry and are often used in more than one MSDS. Putting them in a coded lookup table will save entry time (see Figure 4.6b.)

The biggest problem with MSDSs will arise for the developer. There is no way to predict an optimal number of entries for lists like protective measures, toxicity, and emergency procedures. If respirators are being described, for instance, there could be different respirators for different concentrations of a hazardous chemical, perhaps as many as six or seven in the list. Users should specify this type of occurrence so that the developer can design a solution to the problem.

One last point should be made about MSDSs. There are two kinds, one for the company that manufactures them and one for purchased agents. In industries where this condition exists, users and the analyst should discuss the need to avoid a possible conflict in the system.Manufacturers must print copies of the MSDSs for their products and keep track of where they go in connection with product shipments. Community right-to-know permits anyone to request an MSDS copy from the company, but few people have taken advantage of this opportunity.

Agent Inventory. Two fields describing agents present problems for user consideration. What length of field is necessary for the chemical name of the agent, and how should chemical formulas be written without subscripts and superscripts? Names of some organic compounds are amazingly long. Selecting a field length to accommodate only one or two excessively long names is not practical. It is better to abbreviate the chemical name than to encumber a lot of unused storage space. Writing of formulas will not be difficult. A convention can be adopted to distinguish between subscripts and superscripts.

Record Numbering. Some health and safety professionals have numbering systems for their records, and some clinics number an employee's visits. Industrial hygienists number surveys and samples taken; safety professionals number accident investigations. Consecutive numbering requires special programming. Will the numbering continue to be done by people, or does the user expect the system to supply the numbers? The computer can supply consecutive numbers, and if a record is deleted, that number will not be reusable. Since consecutive numbering will require additional programming, the user needs to weigh the additional cost against the need for this specification. (The cost of many small items may amount to more than the user and developer plan for if many amenities are added.)

4.7 CODED DATA

Reasons for Coded Data. In earlier discussions, the reasons given for coding data were to standardize data and to avoid unnecessary encumbrance of storage space. Both are persuasive reasons, but there is another compelling reason that reflects the basic purpose of the system: to help protect employees from exposures to hazardous substances and conditions.

Job codes track employee assignments, while codes for departments, locations, and processes track physical and chemical agents, as illustrated in Figure 4.7a. Without codes, job and agent data would be difficult to retrieve from the system for evaluation.

From the preemployment medical exam through subsequent periodic surveillance exams and environmental monitoring, there is a constant need for comparison of employee health and safety with workplace hazards in surveilling the workplace. These primary means for detecting job hazards are the responsibility of industrial hygiene and safety. Clinic surveillance is the primary means for detecting changes in employee

```
MODULE: Employee                                         PROGRAM: Assign
                        EMPLOYEE WORK ASSIGNMENT

EMPLOYEE ID: 3456                        NAME: Donald M. Robertson
SOC.SEC.NO.: 222-33-4567                  AKA: Donnie
--------------------------------------------------------------------------
   JOB CODE: 879               DESCRIPTION: Arc Welder, Gas Metal
RELATED JSA: 12-4680 Electrogas welding

LOCATION: Building . . . . 1500
          Department . . .   63 Welding
          Room . . . . . . ____
          Process. . . . .   12 Metal Welding
          Supervisor . . . 4455 Gerald E. Zensky

JOB START DATE: 12/10/1983                                % IN JOB:  60
  JOB END DATE: 05/04/1987                                   SHIFT: 2
```

Fig. 4.7a An illustration of coding job assignment data.

health. A surveillance system attempts to link hazardous exposures to employee illness and injury. The key to this link is a job code that defines work assignments and locations.

Job Codes. Keeping track of employees who are moving through job assignments is a big task for any company. The larger the company, the more demanding the task of keeping payroll records files updated for employees with more than one assignment; time clock cards and differing pay scales must be computerized. In addition, there are changes in supervisors, changes in processing, and absences for illness and personal reasons. Computer record keys can be employee ID number, pay level, and job title, but none of these will describe workplace characteristics.

When the job can be adequately described and related to the hazards present, surveillance can begin. A job code can differentiate among assignment of tasks, locations, departments, and processes. The problem with job titles is that they seldom define job location. For instance, a maintenance person in Department A may work in an environment equivalent to that of a maintenance worker in Department B; however, the exposures may be altogether different. A job code can define more than the pay scale. Job codes should carry some description that becomes a permanent part of that record. They should never be deleted or changed without archiving the original code and description, along with the beginning and end dates of their existence. Otherwise, future analysis of exposures related to that job may be invalidated.

Job safety or job step analyses should be on file for each job code. This is an area where many employers fall behind, but it is an important part of the company's records. Because these analyses match job codes, they should not be continually revised, but retired along with the job code. They should include the agent(s) and condition(s) causing employee stress for each applicable step. This provides the documentation of workplace hazards and is a major bridge between the employee and his or her hazard potential. The first page of a step analysis is shown in Figure 4.7b.

Exposure Groups. Some companies prefer to use zone codes to place the employee in specific locations. Zones are then defined in detail and related to existing hazards. Zoning can be related to grid locations in Title III reporting. Through the use of zones, employees can be homogeneously grouped as to exposures.

Grouping of employees by exposure potential supports surveillance by providing the industrial hygienist with a way to set up surveys and monitoring strategies. Whether the homogeneous groups are by zones or job codes is a matter of company choice based on the recommendations of the industrial hygienist and other safety and health professionals.

For occupational medicine and safety, grouping employees by exposure potential provides a basis for making up examination protocols and training schedules. The clinic knows who to surveil, and how frequently, and what tests to perform. Safety training can be organized around the process controls and protective measures that the employee needs to understand. Some employees will come under more than one protocol because of exposures to several agents or chemical mixtures. In these cases, grouped employees can be covered by special protocols rather than by a protocol for each agent.

Fixed or Flexible Grouping. Fixed groups have an advantage and a disadvantage. The advantage is that retroactive studies can rely on the stability of the homogeneous groups over a period of time when jointly evaluating exposure and medical data. The disadvantage is that processes, controls, and agents may change over time.

Flexible grouping overcomes the disadvantage of fixed grouping so long as every group change is accompanied by starting and ending dates for the group. The disadvantage of flexible grouping is that too much change over a period of time and short-lived groups may not provide enough time to study exposure effects on a small population.

Medical Codes. Medical data of importance in retrospective analysis of individual or group cases should also be coded. Medical data lend themselves well to coding. Physicians and nurses are familiar with codes and abbreviated forms of writing, which can easily be computerized. Data that can be easily coded are shown in Figure 4.7c and an illustration is given in Figure 4.7d.

```
MODULE: Job Safety                                    PROGRAM: JSA
                         JOB STEP ANALYSIS

JSA NO.: 7654  DESCRIPTION: Heat treater entry   REVISION NO.: 2
------------------------------------------------------------------
  BUILDING: 10                                 DATE: 03/28/1988
DEPARTMENT: 45                         APPROVED BY: B. Saunders
      ROOM: ___
   PROCESS: 25 Metal treatment

PPE REQUIRED
     PPE 5678 Disposable coveralls
     PPE 7890 Half mask resp. w/dual filters
     PPE 7654 Gloves, leather

STEPS          SAFE WORK PRACTICE               POTENTIAL HAZARD
  1     Use gloves to carry equipment           Pinched fingers
  2     Get ladder equipment organized          Back strain
  3     Shut down/lockout elevator              Dumping dust
  4     Shut down/lockout air blower            Dumping dust
  5     Blank steam line to treater             Bruised knuckles
  6     Climb ladder to heat treater top        Fall from ladder
```

Fig. 4.7b An illustration of recording a job step analysis in the computer.

MEDICAL DATA FOR CODING

- Reasons for clinic visits
- Medical findings
- Diagnoses
- Reasons for absence
- Procedures
- Medications
- Immunizations
- Body systems and parts
- Symptoms

Fig. 4.7c Medical data that lend themselves to coding.

```
MODULE: Medical                                          PROGRAM: Clinic
                          MEDICAL ALERT

EMPLOYEE ID: 9876                       NAME: Benjamin A. Levy
SOC.SEC.NO.: 333-44-5566                 AKA: Benny
--------------------------------------------------------------------
REMOVABLE DENTURES (Y/N): N             BLOOD TYPE: OO+
WEARS CONTACT LENS (Y/N): N             SEX: M AGE: 20
                   Date       Code               Description
ALLERGIES:         06/17/1983 AA 123    Ragweed________________________
                   08/09/1985 AA 456    Black molds____________________

PROBLEM LIST:      02/02/1986 AP 234    Chronic dermatitis_____________
                   __________ ______    _______________________________

MEDICATIONS:       02/02/1986 AM 876    Aloe cream_____________________
                   __________ ______    _______________________________

INSTRUCTIONS:      Avoid exposures to dermal irritants_________________
                   ____________________________________________________
COMMENTS:
```

Fig. 4.7d An illustration of using coded medical data.

Industrial Hygiene Codes. Hygienists are not as accustomed to coding as medical professionals. This user group may profit by some assistance from the developer in coding their data. Some suggestions are given in Figure 4.7e. The MSDS can undergo a considerable amount of coding, some of which was suggested earlier in this chapter. The systems analyst may suggest additional codes.

Safety Codes. Safety management has employed codes extensively, especially in the area of incident investigation. Most safety data codes have been derived from government mandated reports. Typical safety codes are shown in Figure 4.7f.

For extensive assistance with safety data codes, the reader is referred to the appendices of *Computer Systems for Occupational Safety and Health Management* by Charles W. Ross, (Marcel Dekker, Inc., New York, 1984).

Documenting Codes. Changes to data in form and content over the years keep computer systems personnel alert to the need for dictionary updates. When updating codes, the original code should not be deleted from the system. It should be retired with the date of discontinuance. As new

INDUSTRIAL HYGIENE CODES

Sampling strategies
Sampling types
Sampling methods
Reasons for sampling
Collection media
Analytical methods
Source of reference standards
Exposure types
Agent storage locations
Agent container types and sizes
Agent use frequency
Personal protective equipment types
Survey types
Survey reasons

Fig. 4.7e Industrial hygiene data that lend themselves to coding.

SAFETY CODES

Nature of incident
Frequency potential
Severity potential
Incident type
Unsafe actions
Unsafe conditions
Corrective actions

Fig. 4.7f Safety data that lend themselves to coding.

codes are entered into the system, they should carry a beginning date and authorization. Dictionary descriptions of codes are referenced by the epidemiologist for historical purposes, and any modifications of descriptions will distort future study results.

Codes need to be given to users in a reference manual, as well as being available for viewing on the computer system. Changes or updates to the database must be kept current in the user manual.

4.8 ASSESSMENTS

Assessment is an appropriate beginning in a program for evaluation of employee health in the workplace. Estimates are made of employee health, potential hazards, and the vulnerability of employee health to the hazards. Few estimates are made with certainty, making risk a quintessential element of the assessment process. The fundamental responsibility of occupational health and safety professionals is to reduce the potential risk to employee health.

By a process of qualitative assessment, workplace agents can be ranked for their hazard potential. Once agents are ranked, exposures can be monitored and employees surveyed for the possible effect of exposures on health. Using coded job descriptions and coded agents, agents and their exposure potential can be applied in each of these important tasks:

- Medical surveillance protocols
- Exposure monitoring strategies
- Safety training selection
- Company defense in litigation
- Evidence in compensation cases
- Evidence of compliance with government regulations

Qualitative and Quantitative Assessments. The assessment procedure described below is a qualitative one that serves a useful purpose under difficult conditions. Qualitative assessments are made when insufficient valid data are available and when the industrial hygienist considers them appropriate. Quantitative assessments are possible only when substantial test data exist.

In occupational health and safety, there is a paucity of evidence about the toxicity of most agents, which should not deter health and safety professionals from taking action to control exposures. Whatever evidence exists about the hazard potential of an agent must be evaluated relative to its possible health effects on employees. Therefore, qualitative assessment of the agent and the workplace is a reasonable procedure. As later sampling results prove that the qualitative ranking of an agent needs to be changed, quantitative assessment can be a basis for revision of the rank by the industrial hygienist. Any ranked list of workplace hazard potential based on qualitative assessment should always be subject to revision in the light of new considerations.

Making a Qualitative Assessment. A popular parlor game is to rank things "on a scale of 1 to 10". This is an effort at qualitative assessment based on individual perception of a definition for each rank rather than on a general consensus.

Physicians are accustomed to making risk assessments of laboratory results. They adopt acceptable figures for cutoff points. One doctor may accept a cholesterol level of 200 mg/dl as the high end of a normal range and 201 mg/dl as for the beginning of the high range. Another doctor may choose the normal cutoff as 210 mg/dl and 211 mg/dl as the start of the high range. Still another doctor may choose to make a transition category of 190 to 215 mg/dl. This is professional judgment after evaluating what the doctor believes to be the potential for cholesterol as a health hazard. and the possible risk to the patient's health of making an assessment when there are no guarantees. After all pertinent factors are considered, qualitative assessments provide useful guidance for taking action or not taking action.

Qualitative assessments are useful in depicting a range of conditions, using labels that connote degree even though they have no certain limits. With the addition of numerical ranks for each category, labels are given limits. The number range can be as simple as 0 through 10 with a preferable number of classes limited to 4 or fewer.

Homogeneous Exposure Groups (HEGs). An objective of using a qualitative assessment procedure is to establish groups of employees with similar exposure profiles or distributions, a difficult task since different work assignments expose them to agents with widely varying degrees of exposure levels. A highly subjective procedure such as this should be attempted only by a skilled industrial hygienist.

Figure 4.8 illustrates the stepwise procedure. In Step 1, every job code is listed by process and location; similar work assignments in different locations are given separate job codes. A second list is developed from the first, showing every agent to which employees are exposed at those job code locations. In a third list for each agent, a qualitative assessment is made for each agent by the percentage of working time during which the employee is exposed. A work period can be an hour, a day, or a week, and the employee may have more than one assignment, during only one of which is there an exposure. The percentage of time exposed and the number of classes require a judgment call that the hygienist makes. For

```
1.   For each  JOB CODE
       List  PROCESS/LOCATION
       List  AGENT
       Classify by  % WORK TIME EXPOSED
2.   For each  AGENT
       Classify by  HAZARD POTENTIAL
3.   For each agent in each location and process
       The homogeneous exposure group (HEG) is represented by
       HAZARD POTENTIAL (Digit 1) % WORK TIME EXPOSED (Digit 2)
       If there are three classes for HAZARD POTENTIAL, and
                    four classes for % WORK TIME EXPOSED,
       The HEGs and their rank will be:
                    11     Minor, Minimal
                    12     Minor, Moderate
                    13     Minor, Frequent
                    14     Minor, Maximal
                    21     Moderate, Minimal
                    22     Moderate, Moderate
                    23     Moderate, Frequent
                    24     Moderate, Maximal
                    31     Major, Minimal
                    32     Major, Moderate
                    33     Major, Frequent
                    34     Major, Maximal
```

Fig. 4.8 An illustration of establishing an HEG.

the sake of illustration, four groups are used with arbitrary ranges of <10% (Class 1), 11% through 25% (Class 2), 26% through 50% (Class 3), and >50% (Class 4). Matching word labels will be Minimal, Moderate, Frequent, and Maximal.

Step 2 emphasizes the hygienist's professional judgment in evaluating toxicological and agent data in the context of the workplace, taking into consideration related variables like ventilation and the physics and chemistry of the process.

Toxicological data will include dose-response data in animals and past experience with humans, as found in reference literature. The mode of body entry and the effect of injury must be weighed (such as skin irritant or skin carcinogen, respiratory irritant or respiratory carcinogen), and the amount of exposure time before acute or chronic damage can be expected. While these factors are difficult to classify with certainty, a subjective assessment can be made through the collaboration of the industrial hygienist and the occupational physician or toxicologist. For purposes of illustration, three classes will be selected for assessment of the agent's hazard potential: Minor (Class 1), Moderate (Class 2), and Major (Class 3).

The third step is a ranking procedure that simply combines exposure time classes with hazard potential classes. The combination will show hazard potential (first digit) followed by exposure time (second digit). There are 12 possible results, each one being an HEG.

Using HEGs. What the list of HEG designations says is that groups with major hazard potential job codes at the high end of the range need more surveillance than groups at the lower end of the range. The groups show that resources may need to be deployed most often to higher numbers, with lower numbers receiving less attention. The list can set the stage for monitoring strategies.

In terms of monitoring strategies, previous sampling results may have shown effective hazard control, and a subjective judgment can be made to deploy monitoring resources tentatively to locations of greater concern. These subjective decisions do not disqualify the assessment procedure as ineffective. Indeed, the procedure is effective in that high-priority ranks ensure that major risks were not overlooked, only set aside in the light of other considerations.

Computerizing the Assessment Procedure. A procedure such as qualitative workplace exposure assessment can be designed into the occupational health and safety system. The system does not make the judgments for the industrial hygienist, but design provides the opportunity to make assessments part of the recordkeeping procedures for reference and later revision. A file of job safety analyses would carry the related job code, process code, location code, agent ID numbers, and percentage of work time exposed. Every agent ID record would carry a hazard potential rating. A file of HEGs would carry every related job code, the agent involved, the hazard potential, and the percentage of time exposed. Start and end dates for the existence of the group must be maintained for historical purposes. An HEG file could be a table lookup, depending upon how the system was programmed.

Aside from system design, the system, once in operation, can be used to produce the HEGs by providing lists of job codes, processes, locations, and agents for the hygienist. If a query system is available, the list can contain job codes, processes, and locations sorted by location. Because of the highly subjective nature of the assessment procedure, a lot of manual labor goes into the construction of the original assessments, but once they are entered into the computer, the system can be used cost-effectively. If a company does not have an assessment program, the installation of an occupational health and safety system is a good time to create it.

4.9 IMPORTANCE OF THE EPIDEMIOLOGIST

The Value of an Epidemiologist. Most plants concerned with occupational health and safety do not have a resident epidemiologist; that is no reason to avoid discussing the place of epidemiology in this book. Large corporations employ epidemiologists to observe their numerous plants, and the plants operate more or less autonomously. When a plant automates its occupational health and safety data management, the epidemiologist is infrequently consulted during database design. If this person's services are needed at some future time, data that could have been made useful cannot be used, an unfortunate occurrence that could have been avoided through simple consultation.

Smaller companies have no resident epidemiologist, but it is always possible that the company may find itself in a defensive position when its data may be legally attacked. Having a sound database that has been designed for reliable analytic features may be worth the fee it required for an epidemiologist's opinion.

The Epidemiologist as a User. The role of an epidemiologist as a potential user should not be overlooked during the design of an integrated occupational health and safety system. It is highly desirable to have the advice of this specialist during the design process to avoid potential pitfalls in information management. Although a company may not employ or contract with an epidemiologist at the time the new system is designed, one may be needed in the future. Designing the system toward that end is both discerning and cost-effective.

The assessment program discussed in the last section not only serves the immediate needs of the clinic, the industrial hygienist, and the safety manager, it sets the stage for an epidemiologist in the event that a study is needed. HEGs will be the basis for studying employees who have been exposed to given agents in comparison with those who have not. Direct use will be made of data from:

Demographics	Gender, age
Work histories	Jobs, assignments, start and end dates, locations, agent(s)
Medical laboratory tests	Blood chemistry, urinalysis, special laboratory tests, pulmonary function tests
Clinic visits	Reasons for visits, complaints, symptoms
Industrial hygiene sampling	Results of personal and area sampling for air, noise, etc.

The computer system will provide data for a multitiered epidemiological study. The first tier will determine if the data are suitable for a study. This is why the epidemiologist's suggestions are valuable when designing the system.

The second tier will determine if there is a possible correlation between health effects data and exposure data for suspected agents. Consistency in results is a necessary criterion. Test measurements alone will not support a decision; methods and conditions for testing can influence results, requiring that all the circumstances surrounding the results be examined as well. Text comments attached to data may be very important. In this tier, preliminary statistical tests will be applied before deciding whether to continue with the main body of the study.

Studies. An epidemiological study is a rigorous test of facts which differs from a scientific investigation, as indicated in Figure 4.9.

HEGs have already been selected, but an epidemiologist will closely examine the group for consistency of composition. This becomes a cohort which is consistent for dates and times of exposure to the same agent(s). The examination of employee records across the database helps to determine consistency within the cohort.

Correlating Disease with Exposure. The greatest problem with some agents is the long latency period between exposure and the appearance of disease. This is especially true of a carcinogen like asbestos, which has a latency period of 25 years or more. The shorter the latency period, the greater the likelihood of demonstrating an associative relationship between agent and disease effect.

In many occupational environments, there are multiple factors that may promote disease, either individually or collectively. Diseases of a chronic nature may be difficult to correlate with a specific factor. If the study takes place retroactively, all documentation supporting workplace proce-

DIFFERENCES BETWEEN AN EPIDEMIOLOGICAL STUDY AND A SCIENTIFIC INVESTIGATION	
Epidemiological study	**Scientific investigation**
Non-experimental	Experimental
Observe existing data	Observe new data
Groups established	Groups are selected
Human subjects	Animal or human subjects
Small or large samples	Generally large samples

Fig. 4.9 A comparison of an epidemiological study and a scientific investigation.

dures, recordkeeping procedures, and database composition will be reviewed. This is a forceful argument for conscientious recordkeeping at all times. The epidemiologist does not use all the data in the system directly, but a review establishes the reliability of the data used for statistical evaluation.

Value of Epidemiological Studies. For a company faced with potential litigation from employees, unions, and the public over hazardous exposure effects on human health, the defense begins with recordkeeping and preventive controls. Was a health change caused by an exposure? The first response is to show a qualitative association between the exposure and what may be as undefinable as a nonspecific disease. In view of the liability potential, it is prudent to rely on the expert opinion of epidemiologists. The second response level is to determine a quantitative relationship between the magnitude of the exposure and the degree of change in the human's health. In this endeavor, both the industrial hygienist and the epidemiologist will play a major role. The final response is to show that the possibilities considered in the first two responses are statistically defensible.

There are also less threatening areas for epidemiological evaluation. Epidemiologists can survey the general employee population for incipient health conditions such as minor disorders that are related to hazardous exposure. If this is done, timely corrections can be instituted through additional control measures. Or, if there are early indications of chronic disease, control measures can be reevaluated and changed. This is information is of significant economic importance because of the potential financial drain of disability compensation.

The epidemiologist has a valuable contribution to make to any company occupational health and safety program. Whether the role is full-time or part-time, as an employee or as a consultant, the epidemiologist must rely on the information available in the company's occupational health and safety information database.

4.10 TESTING THE SYSTEM

Quality Assurance. As programmers complete a section of the system, they may ask for data to test with, but most often they use trash data. If the developer has a quality assurance (QA) team, the new system will be tested by them when programming is complete. It is more likely to be the QA team who will ask for real test data. In preparation for this stage of development, users should photocopy filled-in forms or enter real data into forms for the developer's testing team. Programmers test to locate bugs in the programming, which is actually a test for their language syntax and programming logic. In testing the output end of the system, they will also be looking for bugs. The QA team provides the first level of user-type testing. Not being end users with knowledge of the technical data, they are grateful for assistance with real data.

The disadvantage of the QA team is that it consists of systems people, often former programmers, who by training are more likely to be a second level of debuggers. This is not to their discredit; it is only that they cannot approach the system like an end user.

The QA Team. The QA team usually devises a test plan that begins with obtaining appropriate data in sufficient quantity to establish a sample database. The team also wants copies of typical user reports in order to test the output capabilities of the system. The QA team will watch the path that data take from entry to files and from files to screen display or reports. They will know how to track the computer operation through the code, but they will not make code alterations. In Chapter 5, several sections explore the kinds of reports that are needed for occupational health and safety.

Documentation. While testing, the QA team will make notes on suggestions that users should follow in order to make the system work well. This information will be incorporated into both system and user documentation manuals. Programmers have already made notes of their own for inclusion in the system manual.

Test criteria and plans should be documented. System operation can be judged in many different ways, usually depending upon the person's viewpoint. The programmer says that it is a working system if there are no serious bugs; little bugs can be expected and removed as found. The team QA says that it is a working system if the output contains correctly processed data. The developer says that it is a working system if the programmers and QA have tested and passed it, it is run for the users and proves to be error free, and it has everything the user asked for. Users consider it a working system only when it answers all their prayers, which is somewhat unrealistic. This is why testing and accepting the system need to be planned before testing begins. User and developer together should spell out in writing the criteria for accepting a working system.

GIGO (Garbage In, Garbage Out). One of the most confusing experiences users have is receiving a system with a lot of trash test data. It is one thing to read employee records on "Joe Doaks" and "Mary Jane Smith," but it is an entirely different matter to read totally irrelevant test results. Blood chemistry results and sampling results with ridiculous figures tend to create chaos in the user's mind. Even worse is what happens to numerical computations when the input is trash data; the output is so unintelligible that the user does not know whether the system or the data is at fault. If the developer wants to convince the user that this is a good working system, the demonstration should employ real data, preferably from the user's own files, that undergo calculations and come out with the user's own answers.

Kinds of Test Data. Basically, there are four types of data for test purposes:

Live data—data taken directly from the user's files
Real data—valid data not actually taken from the user's files
Erroneous data—data that have been distorted in order to mimic user input error, or data that the system should refuse as invalid
Trash data—An unrealistic grab bag of words, names, and numbers, to be avoided for serious test purposes

Erroneous data are as important as real and live data. Data of this type will be used by the QA team to test processing paths. For instance, if there are normal ranges given for blood chemistry results, will the system tag numbers on either side of the range as abnormal? If sampling results are being compared to reference standards, will the system correctly note exceedance? Data verification can be tested by making deliberate entry errors, such as entering alphabetic characters in a numeric field or entering a zero when the system can only accept a digit other than zero.

Total dependence on live data is not a good practice because most live data lie within the central areas of all possible values. This is why real data to test extremes are needed.

These same test practices should be tried by users, who should test the system in an unhurried manner. Testing is time-consuming, and a large system requires extensive effort. Unfortunately, it is common for users to look quickly at screens full of data entered by the developer. If it looks reasonable, it's okay. This is not testing the system. Users are most often too baffled by what they see and too unsure of themselves to formulate a considered opinion. They need time to assimilate all that is before them.

Taking time each day to examine parts of the system, users should test screens from beginning to end. Use source documents to see how data entry flows, use real data at the extremes of expected ranges, and throw in some erroneous data to test the system's sensitivity.

Developer Responsibilities. Between the time of gathering input specifications for data and completing the system, developers somehow lose concern with data accuracy. Once they know what to put into the system, they focus on solving input problems in order to create a programmable system and place it before the user for acceptance. The loss of interest in accuracy by the time of testing is unfortunate because the game is not over. The user has yet to be convinced that this is a good system.

One of the best ways to "sell" the system is to show the user that it meets the originally described needs. When the user can recognize his or her own data on the screen and in reports, the user is immediately impressed with the machine's capabilities. The user can relate to this kind of review without second thoughts. The QA team should leave meaningful data in the system for demonstration and remove all trash data. What the developer considers to be a "clean" system is a system without bugs that contains everything the user wanted. What the user sees as "clean" is recognizable data being handled in the prescribed manner. Two very different viewpoints!

User Responsibilities. In addition to contributing test data to ensure proper testing of the system, users need to provide less obvious but equally important services to the project. A typical user attitude is that the developer will deliver a product ready to install and use, which is true. However, the user's contribution does not stop with helping to design the system and provide test data; user testing is also necessary. To simply accept a system upon demonstration is very shortsighted. To test a system adequately, users should plan ahead to provide substantial testing time. This does not mean that one large block of time has to be devoted to testing. The written proposal or contract should provide for a period of time in which testing can be carried out in small segments.

Users cannot forego their normal responsibilities to do nothing but test a software system, but they can add to their responsibilities the task of testing over a period of time. Extensive testing will reveal most of the system's quirks, the need for further development, and any additional changes users may need to make in their work practices. User testing is the beginning of system implementation.

Training, Testing, and Implementation. Because the meaning of implementation varies with the individual, it should be spelled out in the proposal or contract. Some people refer only to installation as implementation; to others, implementation includes installation, testing, and training. In a literal sense, implementation means that a plan is being carried out. What the plan covers is the undefined point. In this book implementation includes all activities during the time it takes to install the hardware and software and to establish an operating system that the users can begin to work with.

By the time user testing takes place, the project is traveling in the fast lane. This is downhill time for the developer, who is anxious to close the project, and it is uphill time for the user, who needs to implement the system. The developer is usually anxious to receive payment and move on, and the user will be pressed to accept the system. Smart users will take time to install and implement the system with care, despite the fact that they are being trained, testing the system, and preparing to implement all at the same time.

Ideally, training, testing, and implementation should take place sequentially and unhurriedly, but in the real world that does not happen. Users are anxious to get their hands on the system and can hardly wait to be trained before starting system testing. By the time they start testing, they are already reorganizing their work to implement the computerized occupational health and safety system they have anticipated over a long development period.

In this book, user testing is discussed before user training, but in reality, users cannot test until they have been trained to use the system. The reader should not be disconcerted by the appearance of a contradiction. User testing is included in Chapter 4 because it is a user contribution and is distinct from testing by systems personnel. Implementation and training are joint efforts of users and developers.

4.11 SUMMARY

The occupational health and safety professionals provide dual services to their companies. On the one hand, they give advice on highly specialized technical matters; and on the other hand, they are expected to be attuned to company management, a responsibility for which many have had no training or experience. The primary objective of the computer database for occupational health and safety is to manage technical data, but at the same time, a secondary objective is to include data that will be of value to management.

Government requirements for occupational health and safety record-keeping are extensive. The most efficient way to organize and retrieve data for government reports is to take into account the many types of data required by government regulations as the applications software is designed.

The user's role in system design begins with collection and review of all the forms used to record data. Questionnaires require special evaluation to determine how to store and use the data they hold. Calculations should be written out in detail so that the systems analyst and programmers can correctly interpret their logic and flow for the computer. All data used in the database should be reviewed for their value and for the requirements needed to process them correctly.

From the extensive review of forms and data, database and screen designs will emerge. The technical knowledge of the users establishes the validity and reliability of the data, while the software expertise of the developer establishes the reliability of the system's management of the data. System design is built around programs that are functionally compatible with user work routines.

Special design features may include tickler files and exposure-related selection of employees for medical surveillance and training. Industrial hygiene monitoring strategies may be organized with the aid of the computer system through qualitative assessment procedures.

All data placed in the database require specifications that the user supplies. Standardization of data improves the reliability of data at retrieval and can be provided by coding data. Narrative text should be used only where absolutely necessary, but most textual data can be avoided by coding.

Both users and the developer must test the completed system with live, real, and erroneous data to stress the software and to observe the reliability of the system. Testing protocols should be developed which will provide evaluation standards for acceptance. At the same time that the system is being implemented, user training is taking place and user testing is in progress. Writing of documentation will also be in development, and observations made during testing will provide valuable information for users that should be included.

Chapter 5

USING THE COMPLETED SYSTEM

Following development, implementation is the phase consisting of a combination of user testing, training, and system evaluation. No distinct pattern can be defined because so much depends upon the individual environment. Implementation is not a quick installation process. The integrated occupational health and safety system may take weeks or months to be assimilated into work routines because it is a comprehensive system. User testing was discussed at the end of Chapter 4 because it is a definite user contribution. Obviously, users cannot test a system until it is installed and until they have had some training in using it. This chapter begins with training, which is a definite contribution of the developer, but successful training depends heavily upon the user's cooperation. The same is true of documentation and programmed reports. This chapter focuses on the working use of the system.

Users should be trained by persons who can relate to the safety and health field, and systems personnel should be trained by systems specialists. Documentation includes different types of manuals for operations personnel, system managers, and users. Help screens are a form of on-line documentation that are used for quick reference while using the system.

Output is in several forms and can be considered processed data retrieved from the system as information. When information is organized and interpreted, it becomes knowledge, which can be communicated to others in reports. Reports can be formatted in numerous ways.

Ad hoc reporting is for data review and one-time needs. Using query languages, data can be retrieved from a relational database very efficiently. Repeated data review and analysis is an aid to subsetting data for further statistical analysis before including them in a report.

Profiles are reports that describe a person's health and/or exposure record or some condition in a company or department. Government-mandated reports are expensive to program into the software, but their unchanging nature and long-term usefulness make them a worthwhile user expense.

The developer-user relationship does not end with the completion of implementation. Software repair and maintenance will continue to be required over a long period of time.

The combined resources of industry, health and safety professionals, and government are needed to employ improved computer technology for data management and reporting through an integrated occupational health and safety system.

5.1 IMPLEMENTATION

A New Beginning. Now that the house has been built and it is time to move in, there are many decisions to make, both small and large. This is implementation. Installation of the new computer system is traumatic and chaotic, requiring weeks before smooth operation can be expected. Great flexibility is needed by the entire staff while implementation proceeds one step at a time.

The hardware has to be installed and operating when the software arrives. Software installation begins with the operating system and the DBMS, followed by the occupational health and safety application. The first data have to be imported from payroll for the employee and work history database. This can be done by information services. For a purchased package, the developer will only install the applications software; everything else will be left to information services. Initial installation is relatively straightforward. The implementation that follows will depend on available personnel and money.

If the old database has been computerized, a programmer may be able to recode the data so that they can be automatically transferred to the new system.

Old Data and New Data. Prior to implementation, a critical decision has to be made about old data. If the database begins with the new system, occupational health and safety data will need to accumulate for at least five years before there is sufficient information to begin studying the overall effects of employee exposure, and using all the previous data will speed up the process. Thus, the question is whether to enter previous data into the system. The usual deciding factor is cost; some users just cannot cover the expense of a data entry operator for the purpose of entering old data, and the regular staff will have all it can do to establish a new database using current data. In addition to increased staff requirements to enter old data, there must be machinery to work with; a partial solution is to enter the old data during off-hours.

The advantage of entering old data is that data trends can be followed and data are readily available for any necessary studies. There may be small problems to work out, such as how to make old data fit new forms, but new data requirements will not change so much that old data will be rendered unusable. Some companies consider old data an asset they can ill afford to lose, regardless of the cost. They may retroactively convert data that go back one to five years.

Costs of Implementation. The goal of implementation is to accomplish everything at a minimum cost. Schedules are made for hardware installation or reconfiguration, software installation, office disruption, testing, and training, but schedules will go awry, so margins must be allowed for the unexpected. While this adds expense, there is no alternative. If managers are kept informed of the progress during this period, they can usually find additional resources because they know how much has already been invested in the project.

Two costs that are always underestimated are for testing and beginning operations. While all developers know the odds against perfect testing and initial operations, they never stop hoping. The developer builds in a fudge factor for this part of the project. If users were equally realistic, tensions would be reduced. Testing is a tedious, time-consuming procedure that was discussed in Chapter 4. Following testing, the system is put into action. The wisest method to follow is parallel processing, but not every office may be able to afford the cost.

Parallel processing means that both old and new data systems are operated simultaneously for a planned period, with the old system gradually being phased out. The advantage of this method is that it allows users to test the untried system against the familiar, reliable system. When everyone is satisfied that the new system works effectively, there is confidence that data will not be lost or error prone. The disadvantage of parallel processing is the personnel expense of running simultaneous data management systems.

In one occupational health clinic, the expense of parallel processing was factored in for a period of six months, which seemed excessively long, but the clinic escaped disaster nicely because only a small part of the system ever ran successfully, work was minimally disrupted, and the new system was ultimately rejected by the purchaser. Everyone's high expectations in the computer business *can* founder. This story has a tough message: Planning should be realistic. The middle-of-the-road approach is to allow for a long testing period and be certain of the system before putting it into operation. This requires less time and effort, but it does increase the cost of testing.

There are always incidental costs at implementation for paper supplies, diskettes, and tape packs. Backups require more diskette and tape supplies than were originally anticipated. Check out security arrangements after installation to see if small additional precautions need to be taken like locks on desks or files.

Developer's Schedules. The developer will install the occupational health and safety system by module, rather than all at once, to avoid chaos. All departments will find small bugs during testing and will want some additional modifications. The developer needs time to do this work and can provide better service by bringing departments up to speed one at a time. This gives the first user departments an advantage because the developer will be on-site thereafter and available to help users.

One service that developers are obligated to provide is often behind schedule; manuals should be ready when the system is installed, but typically they arrive late. This makes it harder for the user to train, test, and implement. Users should press for their documentation when it is needed.

User's Schedules. Users should plan flexibly and avoid tight deadlines. Two things to keep in mind are: (1) make hard plans only one day at a time, and (2) plan for daily short staff meetings to discuss the successes and failures of the day before, making additional plans. Office disruption affects not only system users but nonusers as well. For example, if new forms are being used, data entry may need adjustments to complete the forms, and nonusers may have concerns about how to write the forms.

Allow time for users to practice making backups, downloading data from payroll, uploading data for archiving, and similar tasks that are learned skills. Running reports is another skill that needs to be learned before someone expects reports on demand.

Evaluating the System. The implementation period is a time for evaluating the system. Every opportunity to see the system in action will elicit a user's response, favorable or unfavorable. Users should compile a list of problem areas, suggested improvements, and things they like about the system. Evaluation can include items like those in Figure 5.1.

Discuss the list with the developer. Some problems may have rational explanations, while others may be repairable, and still others may be overcome with additional training. The list can be drawn up a little at a time from the end-of-the-day meetings, but it should not become a gripe list. The developer needs both positive and negative feedback. To know that something is working well is to know not to touch that part of the code.

Acceptance of the system depends on the outcome of evaluation. This

```
IMPLEMENTATION EVALUATION CRITERIA

      Developer
        Adherence to schedules
        Helpfulness
        Responsiveness
        Thoroughness

      User-interfaces
        Screen presentation
        User-friendliness

      Data processing
        Error checking
        Response time
        Returning correct lookup values
        Accuracy of calculations
        Text handling

      Training
        Agenda
        Completeness
        Training aids
        Training manual
        Presentation
```

Fig. 5.1 Suggested criteria for evaluating the new computer system at implementation.

is a critical point in the success of the project and signals the fulfillment of the development contract.

5.2 TRAINING

User-Friendliness. Theoretically, the more user friendly the software, the less training is needed. The corollary might be that you can judge the degree of user friendliness by how much training a user requires. Note that the theory does not preclude training; it only speaks to the degree of training needed. Users always need system guidance, whether by formal training or by reading an instruction manual.

User friendly features in a system enhance training and manual instruction by helping to relax the user. Clear, concise screen instructions create a communicative relationship between machine and user. If screen menus and prompts are simple and informative, with easy directions, the user can maintain continuous communication with the machine and the need for training will be reduced.

In reality, a system as large as an integrated occupational health and safety system is never so user friendly that it precludes both training and manual instruction. The extensive programming required to make the system compatible with the specialized needs of three different disciplines requires that the user understand the system.

Training. Several kinds of training will be needed for an occupational health and safety system. Information services personnel will need training for system operation and maintenance. If 4GL tools like a screen painter and a report writer come with the applications software, information services must also be trained to use that software. Training of information services personnel will not be discussed in depth, since this book is a user's guide.

Users, information services, and developers are jointly responsible for training success and must plan together for training because it requires advance preparation. Preparation is a time-consuming task; it will be cost-effective if users and developers work together.

The biggest training cost is for travel, whether the trainees go to the developer's shop or members of the development team travel to the users. An advantage of in-house development is the elimination of travel for training. Under any circumstances, preparation time and training are expensive. Users are away from their desks and have to work overtime or find substitutes to keep up with their assignments.

Materials for Training. Materials for training require advance preparation. These can be training manuals, overhead transparencies, slides, audio cassettes, videotapes, computer diskettes, or prepared handouts. Each type of training aid serves a useful purpose when added to the hands-on experience. Slides and transparencies help the trainee to remember what the trainer is saying, a picture being "better than a thousand words," but when training ends, trainees have no reference materials to rely on. A good trainer will combine screen training aids with handouts showing what is projected on the screen. The trainee can then write notes on the handout for later reference.

Aids that trainees can keep, like cassettes, tapes, diskettes, and training manuals, have a long life and can be used after training as needed. They also help to train new personnel. These aids are costly for the developer to prepare, and for a one-time development the expense is too great. Only when a commercial developer intends to sell many packages are these training aids economical.

Many software packages come with tutorials on diskettes that are of dubious value. The problem is that the user is not being trained to use the intended software system. The tutorial is a program in itself that displays instructions as well as fragments of the application software. Training should be done only on the user's software. In following tutorials, users must read, then perform, read again, perform, and so on, and the process of switching attention back and forth interferes with the learning process. Tutorials on audio cassettes and videotapes are preferable. Audio cassettes emphasize aural learning while the mind concentrates on directions. Videotapes can be distracting if used while practicing on the computer. They are much more effective when viewed separately, and they can be replayed until the trainee is ready to use the machine.

Training manuals, like handouts, help trainees during hands-on training. For group training, several levels of exercises should be prepared to provide additional work for those who learn faster and can proceed on their own while the trainer is helping individuals.

The Trainer. Training aids are one part of the developer's contribution to the system; a good trainer is the other part. A common problem is that good systems designers, programmers, and marketers are not good trainers. Safety managers are more knowledgeable about quality training than software developers because training is a part of their job and they understand how to do it. The developer should provide trainers with knowledge of occupational health and safety because the trainers must relate the application software to the trainee's needs and work practices. User training is not a place for systems-oriented trainers.

Training and the End User (Navickas 1987, page 1) aptly describes the qualifications of an ideal trainer. "The ideal trainer is a training professional who possesses good oral presentation skills, is experienced in instructional design, and is technically proficient in the subject matter". Further, the trainer is advised to develop skills for training and "a training program that meets the needs of the organization". Every training situation differs, although there are always similarities. The trainer has to be adaptable to these varying needs.

Training is not teaching in the sense of providing an education. Education emphasizes the power of reasoning and using one's mind to solve problems. Training is a form of teaching that emphasizes behavior patterns according to set rules and conditions. In education, a problem is open to varied solutions. In training, specific directions are provided for given conditions. The training emphasis is on disciplined behavior patterns, which is why the safety manager trains for correct use of tools, machines, and PPE. Software training should emulate safety training by being specific and disciplined and by providing for the safekeeping of valuable assets.

The trainer is a motivator and sets the tone for enthusiastic acceptance of the system. Computer trainees are frequently afraid of the machines; some are even antagonistic. It takes a lot of hand holding to overcome both fear and antagonism. Therefore, it is important that the developer not crowd the training agenda and provide time for individual assistance. When the trainer overcomes fear and antagonism by gentle persuasion, the trainees are motivated to use the system and the developer is praised for delivering a usable product.

Trainers have to be prepared for any eventuality. Training time is hard to schedule, so trainees come and go during the course of the day; not infrequently, those assigned to be trained make last minute changes in the training schedule. The larger the group, the more difficult it is to give individual attention. The best overall attention is given to the trainer who keeps the trainee's hands on the keyboard. This is achieved by good advance preparation.

What the User can Expect from Training. Training covers the basic procedures needed in applying the computer to daily tasks. The occupational health and safety application is so specialized that training must include individual training for use with work practices. This may entail small procedural changes and accommodation for computer operation. Users who have specific assignments, like being responsible for equipment distribution, maintenance, and calibration, should be given individual attention. Usually, the trainer can suggest easy ways of making work practices and software compatible.

It is probable that follow-up training will be needed because of the amount of detail in the system. The most difficult task for occupational health and safety professionals is to reduce data to codes. For this purpose, the developer must provide training that is relevant to work practices and must make the software applicable to office procedures. Coding is a skill that can be initiated before training and encouraged further during training.

Since codes need to be in place before data input is taught, training should be done in stages. The first stage should focus on understanding the architecture of the system, basic procedures like logon and logoff, backup procedures, and coding of data. The second stage should take place after codes have been entered in the dictionaries, and the emphasis should be on doing data entry. The third stage covers getting reports out of the system. Because so much depends on having dictionaries in good order, the first stage will probably take place early, followed later by the other stages.

A conscientious developer will offer follow-up training beyond the three stages described, giving the users time to become well acquainted with the system. Follow-up training reassures users and eliminates any remaining concerns about the system.

Developer's Advance Preparation. Developers are expected to present a bug-free system, but small bugs invariably creep into the training process. Time for training should be cleared with users well in advance of the training course. This gives users time to organize their responsibilities before training. An agenda should be established with the users because there are always special interests in addition to those tasks which have to be completed. Users should be given a list of what they are expected to do for advance preparation. Prepare training materials to suit the tasks that must be accomplished. Since this is a training setting, materials will serve many training classes, but organization of the materials differs in each environment. For instance, training materials for stand-alone microcomputer users and minicomputer users will differ.

Training materials contain what the trainees need to know. Included in the "what" of software and hardware are terms that may be unfamiliar to the trainees. While trainees will be familiar with technical terms in their occupational health and safety application, the trainer cannot assume that they they will understand how these terms are used in the system, a problem especially with codes. Take the time to put screen prompts in perspective. If the prompt reads "Sampling Strategy" this may refer to either an 8-hour time-weighted average (TWA), or to a method of sampling, like continuous or discontinuous. The emphasis on detail is important, but in a verbal presentation it can be lost or forgotten. If trainees see the detail in the training manual or handouts, they will know where to look later on. The pace should allow input from the trainees and questions as the presentation progresses in order to avoid stumbling points that interfere with learning.

Obviously, trainees need to know the rudiments of using the system, but rudiments have little meaning for trainees with attitude problems or worries about other professional responsibilities, which are common problems in any training situation. In order to keep the hands on the keyboard, have the trainees bring data to class and instruct them to enter them. Gathering data for practice should be part of the user's advance preparation.

User's Advance Preparation. The user's greatest responsibilities in advance preparation are organizing work schedules around the training to provide an atmosphere of concentration for the trainees, and getting the system ready for use by entering codes and dictionary tables.

Organizing work schedules is a real problem for many users with limited personnel and tight budgets. If scheduling is a serious problem, it would be better to schedule individual rather than group training. Groups should be kept to five or fewer persons unless there are more trainers, and training will be easier for everyone if people are grouped by similarity of work assignments. Try not to train groups without a machine for each person. Two people sharing a machine works only if both parties are diligent learners.

Training to use the machine is enough to expect of a trainer; changing attitudes takes additional time and energy. Working to overcome attitude problems begins in development. All potential users should become involved in designing the system, which gives them a participatory attitude. Attitude problems that surface at training time usually arise when users are brought into the project against their will. The trainer cannot be expected to change this situation.

Users should designate one person to coordinate training; this individual can be helpful in resolving concerns that interfere with successful training. Motivation begins with the coordinator, who acts as a facilitator to assist, advise, and recommend to both trainer and trainees. The name of the training game is communication, and the coordinator can tactfully provide both positive and negative feedback to assure both trainees and trainer of successful communication.

Rooms and equipment must be reserved in advance; a schedule should be prepared for the training and work schedules; and the coordinator should make certain that the system is ready to use when training begins. An integrated occupational health and safety system such as the one described in this book requires preparation before use. Codes and lookup tables need to be in place before screens will operate. Testing calculations with live data in the company of the trainer is a good idea outside of class. These are user responsibilities. Developers are expected to produce an operating, bug-free system. Information services must get codes and lookup tables in place before expecting people to enter data.

Systems Personnel. Information services should provide basic computer training for those who need it, which includes computer basics like booting up and closing down the machine, logging in and out, making backups, keyboarding, and system architecture. This is not a responsibility of the occupational health and safety trainer. Lacking in-house assistance, there are appropriate training classes everywhere in adult and continuing education courses, as well as companies whose business is computer training.

It is helpful to include someone from information services as an observer in the user training class. This can be someone assigned to the occupational health and safety system, the database manager, or the systems administrator. While this person will have already received systems training in the software, he or she can assist the trainer with individuals, become acquainted with some of the user's concerns, and provide both trainer and trainees with information about the system's environment. The database manager or systems administrator will be in continuous contact with users, making this a good time to begin communicating.

5.3 DOCUMENTATION

The Importance of Documentation. Documentation is an essential part of the software package that describes system architecture and system use. It serves as a reference guide for systems personnel and users. Not only will it be used to operate the system, but it will refresh memories and help diagnose errors when something does not work. This makes it an indispensable part of the package.

Documentation is the written form of communication about the system, and training is the verbal form. Verbal communication is effective in the short term, but written communication is made for the long term. Documentation is by far the more difficult medium because the style is formal and must satisfy users with all degrees of computer literacy and perceptions. The best documentation is written in a personal style. When training is over, documentation will continue to teach the skills and disciplined behavior that make the software work.

Kinds of Documentation. An operations manual is written in the technical language of computer systems for information services personnel like computer operators and systems management. Among other things, it describes operating commands and procedures, installation and startup instructions, file recovery procedures, and backup procedures.

The system or application manual is directed at the system's management and describes the system's architecture, functions, and capabilities. It is a database administrator's guide and uses language familiar to database personnel.

For users there are two types of helpful manuals. The first is a training manual that explains the use of the system in a stepwise manner. After discussing the basic architecture and functions of the system, general skills are described, followed by special functions and the skills needed to use them. There are many illustrations, usually representations of screens. General skills are procedures such as logon and logoff, using menus, and entering and retrieving data. Special skills include navigating around the system, performing calculations, and running batch operations. Exporting data to interfacing systems for statistical analysis is another special skill.

The user's manual is the second type of manual written for users. It is written clearly and concisely to describe functions like adding, changing, and deleting instructions for data entry, browsing through files, and most of the topics covered in the several levels of training. Error messages are listed and should carry instructions for overcoming a temporary halt in operations. This becomes the user's companion when using the system.

Depending upon the application, report writing may be described as part of the system or discussed separately as a function of the report writer. While report writing, using SQL, and exporting data are skills that must be acquired, they are more likely to be covered in follow-up training and may be described in separate manuals.

Another kind of documentation may come as a manual, or it may be embedded in the code. This is documentation for programmers and is a necessity for maintenance and repair of the code. It is used by programmers and is written in a technical language they can understand. In includes system flowcharts, a detailed description of the system modules, file information, and code commands. Instead of being only a manual, it can be the source code for the package, but this depends on the terms of the contract. Developers do not always release source code.

Finally, screen painters and report writers will carry their own separate documentation. Since these 4GL tools are used by information services personnel, the documentation will not be for users.

Evaluating Documentation. The typical user reaction to a computer system is that the system is good but the documentation is either "so-so" or "awful." The system is not characterized as "terrific" because, however good the system is, poor documentation degrades it. Covington (1985, page 165) makes the provocative statement that "documentation is arguably more important than programming . . . the program becomes usable only through documentation." He cautions that the objective is not to show off mastery of elegant language but to "get your ideas into the reader's head as easily as possible (page 168)." Lastly, he wisely observes that "you don't release untested programs; why release untested documentation (page 172)?"

A systematic evaluation of documentation depends on the system's being in operation. The question is how easily the system can be used by reading documentation. Criteria for evaluation are shown in Figure 5.3.

This list is not intended to cover all possibilities; it only suggests points

DOCUMENTATION CRITERIA

Clarity
Simplicity
Narrative style
Logical organization
Content and completeness
Audience level
Objectivity
Format and appearance
Length
Headings
Index
Appendices
Error messages defined
Trouble-shooting instructions

Fig. 5.3 Suggested criteria for evaluating computer documentation.

to consider. Three items, however, should be stressed. A good index is imperative for documentation, with ample cross referencing of topics. Users are easily turned off when they cannot find what they need to know, and this is the point at which they consign documentation to the ash heap.

The second imperative is that the subject matter be broken down into short topics with quickly identifiable headings so that users do not have to read five pages of text before they find what they are in a hurry to know. This can be overcome by using an accessible format. Headings in boldface print or marginal annotations are quick ways to find what the user wants.

The third important inclusion is the appendices which can list function keys, error messages, special commands, a glossary, and similar supportive information that is too voluminous to incorporate into narrative text. Frequently, these supplements are referenced by users more than any other part of the manual. As an accessory to the documentation, a "cheat sheet" to keep next to the keyboard is the way to a user's heart. Depending upon what the sheet is intended for, it can be called a *keyboard chart*, a *function key card*, a *command card*, or something similar. It is helpful to have this printed on heavy stock for wear and tear. Computer stores carry templates to place over the keyboard for well-known packages, but this is too expensive for the occupational health and safety developer. However, it should not be too difficult to print a single sheet which will provide support for beginning computer users.

The greater the number of users for the occupational health and safety system, the more important it is to have good documentation. Consider documentation as the user's "right-to-know." The real test of documentation can occur only after the system has been used for a time. If this has truly been a user's handbook, it will have been taken from the shelf often, with dog-eared page corners and numerous coffee stains.

Other Documentation. The developer is not the only party responsible for documentation. Systems management, the database manager, and the safety and health departments must put together procedure manuals for their own use. Reference manuals for files, variables, and codes need to be written and kept updated.

Help screens are on-line documentation for the user's convenience during system use. This is discussed in Section 5.4.

Security for Documentation. Documentation tends to be treated casually by users, which presents a problem for system security. Manuals contain explicit directions for operating the system, and reference books contain lists of files and variables. Locking up machinery is the first line of defense, but some thought should be given to preventing system misuse and file tampering by protecting documentation against unauthorized use.

5.4 HELP SCREENS

Help Screen Documentation. Help screens are an important form of user documentation that is provided by the developer. Unfortunately, they lack the design assistance of users. Although the developer uses the help screen feature to sell a system, the feature is thrown in as a last-minute task before user testing. The result is that users do not find help screens of any great value, which is a pity because they can offer on-line assistance to an impatient user.

Help screens are intended to be a quick source of information during system use. They should communicate to the user what the system is expecting and can be as detailed as the developer can make them under the limitations of time and memory capacity. Help screens are not a tutorial, nor are they designed to replace user manuals. The design objective should be for the user to go directly from the part of a screen that presents a problem to the documentation that will overcome the problem and return the user to the point of initiation.

This is a sizable task for the developer for several reasons. The first task is to anticipate when the user will experience problems. The systems analyst and programmers will identify some points, but generally it is the user who uncovers problems at testing and training time. Help screen development usually takes place after system development when the developer has conflicting interests, namely, to deliver the system to the users for testing at the same time that help screens are being added. Reaching a decision about the optimal amount of on-line help that the user will need and how much memory space can be allotted for help screens is another problem for the developer. Lastly, the developer may already have a cost overrun, and help screens only pile on more programmer time.

Designing the Help Screen. Help screens are drawn from a secondary database that contains all the necessary information. The screen display can be full screen or windowed. Either one should have the ability to scroll up and down. Help screens are entered by the use of a function key, and the Escape Key (ESC) is normally used to return the user to the original screen.

Help screens are often designed for use by levels in menu-like fashion. In a multilevel help screen, it is helpful to highlight the options one at a time and let the user make a selection with the Return Key. However, some developers require that the user type in the preferred option (Figure 5.4). The first level might give a list of topics like Files, Keys, and Variables. If Variables were selected, a second level might read "Field Variable, Default Variable, Computed Variable." Selection of an option would lead to a glossary-type definition of that kind of variable. A more

```
     For Help with the Safety Investigation Program enter
one of the following topics:
  Investigation   Causes   Corrective Actions

TOPIC> Investigation

     The Accident Investigation form records data about a
work-related accident.  For help with the form enter one of
the following subtopics:
  Place   Witnesses   Personal Injury   Property Damage

SUBTOPIC> Property Damage

     The Property Damage section records the details of any
property damage because of the accident.  Selection of the
Vehicle option will jump the user to the Vehicle form for
additional details.  Press Ctrl-C for a list of codes to be
used in describing Property Type and Equipment Type.  Source
and Extent of Damage are also coded replies.  Use Ctrl-C for
these codes also.  The Comment field can be used for a brief
narrative for further description of the damage.
```

Fig. 5.4 An illustration of a multilevel help screen.

sophisticated help screen would go directly from the variable the user was on when help was requested to describe that variable and what the computer expects to have entered. This is called *context-sensitive* help. Using "Default Variable" as an example, the help screen might say:

> Start Date—a default value that uses system date. Change as desired. Accepts digits only. Slashes permanent. Examples: 02/08/89, 12/10/89.

It should be pointed out that a description of this length would present memory problems with the hundreds of variables in an integrated occupational health and safety system, especially for a microcomputer. When developers attempt to solve the problem, they invariably look for ways to abbreviate information and find ingenious ways to say a lot in little space. If they did this with the guidance of the users, users could give advice on formats that are understandable to them. Ironically, programmers devise intricate patterns for help screens that frighten users the first time they use on-line help, thus defeating the purpose of providing help screens. Help screen writers should not be programmers, and the help instructions should be succinct and follow a consistent format.

Types of Help. Help can be used for keyboards, files and variables, and the purpose of the screen. Providing the user with function key descriptions is a common keyboard help aid. These keys need to be described on-line because they replace repetitive tasks, saving user time and energy. Some keyboards contain more than one kind of function key. For additional tasks, combinations of keys are programmed, using keys like Control (CTRL), Alternate (ALT), and Shift in combination with function keys, number keys, or alphabetic keys. With so many possible combinations and actions to be programmed, on-line help is mandatory. It should be pointed out that microcomputer keyboards typically have the Alternate Key but larger machines do not. Keystroke combinations on the larger machines use the Control key, function keys, and number pad keys. Some keyboards on larger machines have additional special function keys.

Files and variables are commonly described by help screens. The default variable shown in the discussion on multilevel help could be used in a screen describing a file and its variables. The purpose of the file would be briefly cited, and each variable would be described in the order of its use on the screen. The type of data expected, whether or not it is a key, and the length of the field would be given. It is helpful to do this in narrative style for the user's benefit, as shown in the example.

5.5 INFORMATION

Communicating with Management. A persistent subject of papers in occupational health and safety journals is how to communicate health and safety information to management. One of the goals of installing a computer system is to improve reporting to management, the underlying myth being that the computer will be a fairy godmother who magically delivers a pumpkin coach to management's ball, with happiness ever after. The reality is that the computer is not magical; it is a mundane machine that slavishly does what humans direct it to do. It is humans alone who can impart information to and communicate with management. Using the computer as a tool for this purpose requires that users make the effort to determine what needs to be done before asking the computer for help, enabling the computer to perform to the best of its ability.

Information and Knowledge. The computer can only accept data, process them, and return them in processed form. Processing of data involves a conversion of the original input form into another serviceable form. The computer will mechanically process a piece of data for screen viewing, storage, or output. If given specific directions, it can systematically act on the data by transforming, repositioning, or performing calculations with them. The result of this action, or series of actions, is information.

Information is what the user receives from the computer. This is the organized data. It is factual and meaningless when it emerges from the computer machine. The only organization it has is that which the user instructed the computer to carry out. If there are many pieces of data that the user wants the computer to return, the user can specify that the presentation be in a given format, which simply organizes the data for viewing. The output presentation form is a report that contains organized data.

In the purest sense, information is data organized on a page to be read. This stimulates the human mind to see meaningful and useful relationships. It is the human who transforms meaningless data into meaningful data. A well-organized report can stimulate the reasoning process to move beyond simple comprehension to imaginative levels and to recognize the value of the data in making or proving a point. The role of the computer is to process the data for presentation as directed. The role of the human is to determine how the data need to be processed and presented for comprehension. If the human errs in giving directions for either processing or presentation, comprehension is diminished.

Information is a material commodity produced by machinery under human direction, and it can reach loftier levels of importance if it imparts knowledge. Knowledge is comprehension and understanding of an organized body of information. It is not a material thing like information; it is the product of the brain that allows the human to have an idea and to act.

Sir Francis Bacon said that knowledge is power, which brings us to the subject of management. A common lament of occupational health and safety professionals is that they have no power in the management structure. If Bacon is right, then professionals lack power because they do not use knowledge correctly. This is not a specious argument; it is an empirical observation of the user of occupational health and safety systems. Health and safety professionals must analyze the place of the computer in their company roles and their relationship to management. The kinds of reports they want and the design of reports are dependent on the role that they expect the computer to play. Will the computer produce reports with usable information, and will this information lead to knowledge? Simply regurgitating data from the database will do neither, and yet that is what most occupational health and safety computer reports appear to do.

The best definition of the roles that computers play in systems like health and safety data management was given by Mervat Abdelhak (1980). He says that there are two types of models for health systems records: the Data System model and the Information User System model. "The decision of whether to have a Data System or an Information System should be made based upon users' needs, objectives, expectations, and the available resources" (page 107). The Data System is a logistical system that is primarily for collection and processing of data when information resides in the system and is regurgitated upon demand. It stores and produces information in an efficient and effective manner. Data is a commodity that can be "retrieved, purchased, packaged, and transferred" (page 105).

The Information User System emphasizes formulating knowledge from data input and processing, the "process of data transformation followed by user transformation that leads to change of the state of being" (page 106). He concludes that "the information user has an integrated knowledge of the state of events and subsequently feels informed" (page 106). If occupational health and safety professionals are truly concerned about their relationship to management, they will have to orient their systems toward the Information User model, but if their objective is only a repository for data, the Data System model is appropriate.

> The deficiency of the simple data processing system is that it fails to tell management all sides of the story. . . . That brings up the major

> problem of building an integrated data system: cost. The cost of the system is due to the need for a thorough understanding of all the interrelationships that exist in the business. (Aron 1969, page 218)

If Aron were to describe the data system, he would claim that it was a system for the entry operator, not the professional, since data interpretation occurs at a very low level. The integrated system he considers necessary to support management is the equivalent of the Information User System, which is a computer tool to provide information. He advocates that a fundamental education program be employed to "teach the user of an information system how to get useful information out of the data base" (page 230).

Communication. A primary ingredient in the management process is communication, another long-standing theme in occupational health and safety literature. Communication is the imparting or transmitting of knowledge. Some people communicate more naturally than others, but all can improve their communication skills. Bellanca (1988) wisely counsels that even though all your hard earned knowledge and experience are useful, "your best and brightest ideas are worthwhile only when you can influence, inspire, or discover the motivation which will activate others to adopt your way of thinking" (page A-692). He suggests that "(w)hen you recommend change to improve the healthfulness of the workplace and it is not accomplished, perhaps it is not the result of what you communicated, but how you communicated it" (page A-692).

John Pendergrass sums up the need for good communications by saying:

> Communications keeps things moving. Accurate, concise, understandable and complete reports delivered so that our clients are not taken by surprise will keep people requesting industrial hygiene service. Requested service is a mark of the successful program and of the successful manager." (Pendergrass 1983, page 315)

The clients he refers to are executives, line management, other departments, unions, the public, and governments, which covers almost all the possibilities for occupational health and safety reporting. The occupational health and safety computer system will provide the information to be imparted to these "clients," but it is up to the professional to determine what needs to be communicated and in what form.

A Communication System. A simple communication system is shown in Figure 5.5a. Information is encoded for transmission through a channel to

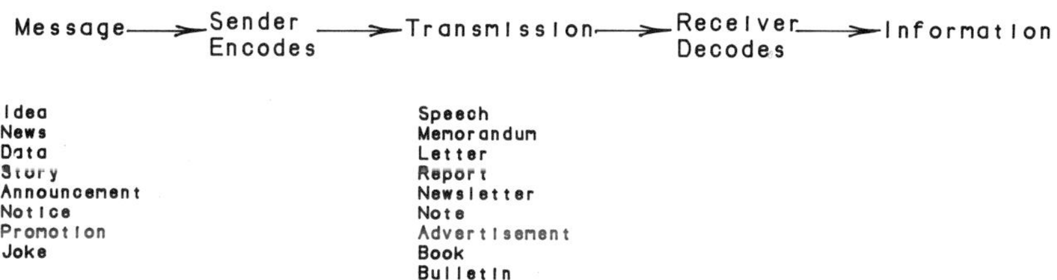

Fig. 5.5a An illustration of a simple communications system.

a receiver, who decodes the information for assimilation. This process meets with interference at several points. At the encoding station, the sender has to select terms that the receiver will be able to understand or decoding will be impossible. Special terms need to be sent with their own decoding information to avoid decoding errors.

The proper transmission channel has to be chosen, and quality transmission is a goal. If the channel is paper, there must be good quality print, good margins, understandable headings, and easy-to-read, clean copy

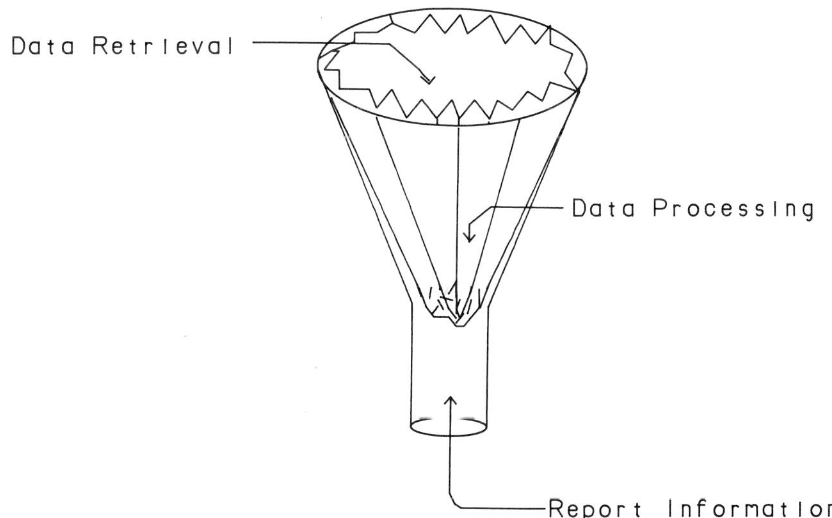

Fig. 5.5b Processing data in a computer for report generation.

with no spelling errors. If the channel is sound, the presentation needs to be clear, concise, and easily understood.

At the decoding point, there must be no psychological or physical interference. The decoder is aided by familiar connotations and terms. Good connotations enhance decoding, and bad connotations disrupt it.

If the contents of the communications to management must be technical, keep the information simple and direct. Avoid a lot of data, and present summary data rather than quantities of test results. If the data can be correlated with dollars and cents or production figures, more information can be given, but the volume needs to be controlled.

Preparation for communication with management should be like the analogy shown in Figure 5.5b. Pour the data into a filter funnel. The filter will screen out everything but that which is pertinent. (The computer is an excellent filter tool!) Coming out of the spout will be a small quantity of special data, and left behind in the inverted cone is the unnecessary data. The way to prepare a report is to evaluate the information that the computer has to offer, decide what filter will produce the most desirable output, and adjust the appearance and communication form to the intended audience. One report need not tell all. Information left in the funnel can be filtered again for a different purpose. Several reports may communicate better than one, possibly spaced over time.

Principles of Computer Report Writing. The objective of this section has been to present important principles for use with the occupational health and safety computer system in providing information and knowledge. A common method of using information is to share it with others through printed reports, but the system can be used in two ways: to output reports and to help the user review the data before deciding what data represents and how it should be reported. The review process can lead to data analysis before reporting. (See Section 5.9.)

Since users must be encouraged to consider what the computer and the occupational health and safety application software can offer, this section has explored the fundamental relationship between the computer system and the human. Hopefully, this will not be seen as an esoteric exercise. The user needs to ponder the principles of the Data System vs. the Information User System and to decide which one suits the user's requirements. If the Information User system is chosen, the transition from data to information to knowledge, and the means of communication, need to be explored.

As a final thought, it is worthwhile to note that wisdom is knowledge of facts that are assimilated to provide insight and good judgment about people, life, and conduct. It is wisdom that health and safety professionals hope to acquire after reading occupational health and safety computer reports. Above all else, it is wisdom that the health and safety professionals must exercise if they are to justify this expensive computer system.

5.6 OUTPUT AND REPORTS

Occupational Health and Safety Information. Data retrieval from the occupational health and safety database can be varied according to the needs of the users. Modular construction of the database lends itself not only to the varying data input requirements of the different health and safety departments but to their report output as well. Each module processes its own data, performs its own specialized functions, and can act as a standalone system if necessary.

An important concern about the database is to avoid redundancy. Reporting in any form reinforces the necessity to guard against redundancy because the database must not output the same fact twice. Every item in the database contributes to information of some kind. The vital need for consistent data, stressed in an earlier chapter, is important at data retrieval time.

Types of Output. Computer information is presented by two methods: on-line screen display and peripheral hardware like printers and plotters. A third type, a robotic computer voice, is still in experimental stages.

Screen display is on-line information retrieval and is readily available to the user. Two types of file data can be viewed: The first is a request to see a specific record; the second is a wild card request to review a group of records, which could be all the records in the file, one at a time, or all those matching a designated sequence (string) of characters. A typical request is to give every record in a specified file for someone with the last name of "Stephens," displaying them one at a time. In some systems, wild cards work only on key fields, while in others, they work on any field.

On-line viewing of the database, screen by screen, is called *browsing*. Editing of the database can also take place, which makes browsing especially useful. It differs from the normal screen viewing in that the user can view record after record without resorting to specific record retrieval. This works well for MSDSs.

Browsing suits the unsophisticated user who is not familiar with command language and needs only to use a few function keys to move around the system. For the professional user, browsing provides the opportunity to review all records pertaining to one or more specified fields.

In addition to on-line screen display, information can be retrieved by batch processing. During peak use, minicomputers and mainframes cannot handle the user load, creating a slow response time. Rather than add to the burden, some output can be programmed for batch processing. By batching together several or more complex retrievals to run at a low priority, the files can be searched and data processed at off-hours without interfering with other work. Batch retrievals are frequently done overnight. Examples of searches that lend themselves to batch processing are running tickler files and printing out profile reports.

Reports for Communication. Reports are a means of communication that provide a flow of reliable information for making decisions. Douglas (1957, 3) says that "a good report puts hard thinking into plain words for quick reading to guide prudent action." Citing the origin of the word reports as coming from the Latin *reportare* (to bear or bring back), he describes the reporting process as going out to look for information, researching it, bringing back the facts, explaining what the facts mean, and recommending action. When the reader understands the situation, he or she makes a decision, action is initiated, and control takes place. Douglas claims that reports are "a mental and linguistic performance of a high order," requiring mental discipline. The ability to write reports is a fundamental skill for a manager. Competent report drafting requires:

- Careful inquiry technique
- Orderly progression from problem to solution
- Capacity for exactness
- Attractive report appearance

Occupational Health and Safety Reports. Occupational health and safety reports serve many purposes, and each type of report meets a different objective. Because each company has its own names for reports, a brief description of the many types of reports will be given:

- Standard reports—provide information on a regular basis, like monthly listings of laboratory results.
- Exception reports—lists of deviant data such as employees who were scheduled for medical tests and failed to show or test results that lie outside of acceptable range.
- Production reports—usually periodic, designed to keep track of administrative information like how many employees were seen in the clinic in the past month, how many sampling surveys were conducted, or how many incident investigations were made.

ABC Corporation July 1, 1986

QUARTERLY INDUSTRIAL HYGIENE REPORT
Second Quarter

	Number Performed	Within Limits	> 1% Exceedance	Test Cost	Total Cost
Compliance Testing					
Volatile Organic Hydrocarbons	5	4	1*	$ 45	$ 225
Noise	3	3		30	90
Monitoring					
Volatile Organic Hydrocarbons	8	7	1*	45	360
Asbestos	1	1			
Microscopy				50	
Ultramicroscopy				500	550
Film Badges					
Ionizing Radiation	105	105			105
			Quarterly Cost Total		$1330

*See report for recommendations

Fig. 5.6a An illustration of a management report produced by a computer.

- Special reports—in an ad hoc or formal presentation, these contain specific information needed on a one-time or infrequent basis.
- Management reports—may be periodic, containing costs, personnel information, and production data, but not technical data (see Figure 5.6a.).
- Profiles—can be done on employees or work locations, and may cover the state of an employee's health, as shown by a list of test results over a period of time. Industrial hygiene may show all the assigned exposures an employee has received over a period of time or

the exposures assigned to a work location. The safety profile may include all the incidents in a work area and information about the employee.

- Notifications—form letters notifying employees of medical or exposure monitoring results.
- Data dumps—on an ad hoc basis, these simply give the user all of the data in the database on a requested field or fields.
- Government reports—mandated, usually annual, with a due date and a prescribed format, like OSHA 200 and 101, EPA Title III reports, and Mine Safety Administration reports. State and local reports are also in this category.

Good Computer Reports. Reports are a presentation of information. A well-designed report can favorably influence the receiver even when the information is unwelcome. Therefore, it is important that a report, regardless of its type, be carefully designed for quick recognition of its contents. Some characteristics of good report design are listed in Figure 5.6b.

If the report goes to management, advance preparation should define the content and the format most likely to catch a manager's eye because management recognizes competence by the reports they receive. Analyses of the impact on production, sales, and the bottom line are guaranteed to please. Throwing in large chunks of the database has the undesirable effect of turning managers off.

Reports are easiest to read when there is a lot of white space and a limited amount of information on a page. Reports crowded with columns from left to right are difficult to absorb for anyone but the person(s) familiar with all of the information. It is better to break information down into several pages than to overcrowd a single page.

Computer reports are effectively programmed for plant occupational health and safety by dividing information by department, location, or process. Control breaks cleanly separate the divisions horizontally. A control break occurs when the records in a field shift to the next department (process or location) until all of the records in a department have been output; then a double or triple space is added, and the report on a new department begins. The computer can total within the divisions and provide grand totals at the conclusion of the report. The same is true of making counts, such as for safety incidents and exposure monitoring. Incident costs and counts should be used in management reports. Details of incident information should be separated into reports for safety committees, and exposure results may be shared with safety management, but company management has no use for incident details or exposure results. Report design is a skill and an art.

GOOD REPORT DESIGN FEATURES

Logical arrangement
Concise statement
Visual simplicity
Uniform style
Well focused subject
Directed at a specific audience
Tables organized for clarity and meaningful relationships
Charts designed to relate information at a glance
Clean copy

Fig. 5.6b Suggested features of good report design.

Tables. Tabular presentation is the most common format for occupational health and safety data. Tables quickly reveal the organization, classification, and relationships of the information offered by formatting data in a two-dimensional array. Relational databases are designed to present data in tabular rows and columns.

Some DBMSs offer cross-tabulations, which are especially adaptable to occupational health and safety review and analysis. Given the data groups for the rows and columns, the software sets up the table and enters into the cells counts and percentages for the specified groups.

Stubs are headings that describe the contents of columns and box heads describe the contents of rows, but they are not considered part of the table itself. The body of the table is all the relations combined. Each relational intersection is called a *field* or *cell*. The table must have a title and may also carry footnotes, but these are not part of the table.

Advantages of Good Computer Reports. Despite the time and effort it takes to design computerized reports, there are advantages that make the effort worthwhile, as shown in Figure 5.6c.

5.7 AD HOC REPORTING

Ad Hoc. *Ad hoc* means that which is done for a specific purpose or end. Ad hoc output of data is often needed for a specific review purpose. Typical ad hoc reports concern:

- All employees whose cholesterol level is elevated and whose weight is more than 15 pounds over their ideal weight
- All employees who have been exposed to a given chemical for more than two hours a day over the past year and who are current smokers
- All employees working under the direction of a given supervisor in the past year who have had accidents of any kind

These may not be continuing requests for data, but they are useful on a one-time or occasional basis. On a one-time or infrequent use basis, and without advance notice, information services cannot allocate programmer time to code reports, making query language a very useful tool in the occupational health and safety computer system.

Query Language. Query language is a high-level computer language for the retrieval and modification of data and is ideal for ad hoc reporting in an interactive, on-line capacity. Processing includes simple computations on the data, data sorting according to the user's directions, and data selection for subsetting to new files. In the hands of skilled users, query language is a powerful retrieval tool.

It is not necessary for the users to be programmers because query

```
ADVANTAGES OF GOOD REPORTS

INCREASE IN

     Decision-making efficiency
     Control over costs
     Service to management
     Control over equipment resources
     Control over personnel resources

REDUCTION IN

     Clerical hiring

FOCUSES ATTENTION ON

     Trouble spots
     Goals for improvement
```

Fig. 5.6c The advantages of designing good computer reports.

languages contain very simple constructs of one or two words that fit together in an uncomplicated order. Constructs are programming statements that specify what the computer is to do. Each programming language has its own constructs, and query languages have great power in that directions are given in a minimum number of words. All the user needs to employ query language is a little time and patience to attain a moderate level of proficiency. The value of query language is the flexibility that it offers for extracting information from the database in a multitude of combinations on demand.

The query prompts a database search that selects all data complying with the request. This becomes a database subset that is displayed for the user in tabular format. In essence, this is a report. Query languages are usually part of the DBMS, and many are based on SQL, which was discussed in Chapter 2.

Query languages have their own identity. Queries can be carried out through interfaces to typical programming languages like Pascal and C. Procedural query language specifies how the data are to be retrieved and output, and requires a programmer's expertise to use the correct syntax. Nonprocedural queries do not specify the retrieval or output techniques. SQL is a nonprocedural query language. It specifies only what data are to be found, leaving the DBMS to direct the retrieval procedure.

When nonprocedural query languages are used, the developer has to review the database design carefully to make certain that the relationships between data can be read rather than searching through individual files and fields. This is why relational databases and nonprocedural query languages are typically linked, while procedural query languages are associated with hierarchical and network databases. However, there are some relational databases with procedural languages.

The DBMS's search of many paths in a relational database is supported by the use of an optimizer. This is a component of the DBMS; among its tasks is the provision of optimal paths through the database for search and retrieval. It is activated when the user logs into the query package and remains effective throughout the session. This removes the necessity for reestablishing paths for each query and optimizes the efficiency of the DBMS. An optimizer is not mandatory, but it increases the efficiency of data retrieval.

Query languages evolved with the development of relational databases and the concept of records as sets of data elements (attributes) belonging to a key field. When users take the time to learn query language for data retrieval, the application software becomes more user friendly, inasmuch as the user is maximizing the capabilities of the system.

The Users and Ad Hoc Reporting. Every kind of health and safety user will have a different emphasis in using ad hoc reports and learning how to produce them. The data entry operator uses the occupational health and safety system daily and has a comprehensive grasp of the files and fields in the database, the files being related to the screens, and the fields being related to the prompts. The query language may not be learned, but dedicated entry operators are often persistent people. Since query language is usually programmed to accept commands one line at a time upon being prompted, these users can learn to set up simple ad hoc reports.

The occupational health and safety professional will probably use the system for ad hoc reporting on an occasional basis but will ask complex queries of the database. Learning the files and fields will be difficult because of lack of daily use of the system. A reference manual is necessary to overcome this drawback. More query constructs must be learned than are required of the entry operator in order to use the full range of capabilities of the query language. When information is generated for management, report summaries are given rather than technical data. The range of uses will be limited only by the user's willingness to learn query language commands.

It should be noted that security remains intact for query retrievals. Where users are forbidden to access data, these restrictions continue to be in effect.

Query Output. Although most ad hoc reports are viewed on the display screen, users invariably need to save or print the report for further study. The report can be named and saved in a file of its own for reuse in the future, making it a standard report that can be sent to the printer. Some query language commands will also construct a format for the report as well, providing column headings, counts, subtotals, and totals. The column headings and the report title can be whatever the user chooses. Ad hoc reports are simple and informal, but they have surprisingly flexible formats. When reports are saved and used anew, the data are updated by the system.

5.8 REPORT WRITERS

A Database Tool. A report writer is a 4GL tool designed to output data according to the user's criteria for content and format, automatically generating source code for compilation without the aid of a programmer. It is mechanical and usually provides a large variety of options for formatting the report. For occupational health and safety professionals the amount of detail may be more than patience will bear, and the task of report writing may be delegated to information services.

Reports can be displayed before outputting to the printer, saved for continued use, and modified as needed. They are an enormous asset to an integrated occupational health and safety system because they save programming costs and can be modified quickly upon request. Some report writers are easier to use than others, with menus and function keys providing the options for selection. It is the heavy reliance on function keys and numerous options that persuades users to let information services define reports, even though a skilled user can probably format an report in an hour, which is a cost- effective reason for acquiring the skill.

Formatting a Report. Data for the report are defined by files or fields from more than one file. A title is specified, and a template or framework is constructed for positioning the elements as the user dictates. The width of a report will be limited by the maximum width of the printer line, but the length can be as long as required through paging. When the report exceeds the screen size, the report writer will permit scrolling to the left and right, as well as up and down pages.

Data can be organized on the page according to the user's specifications. Requests for sorting and filtering of the data become part of the report specifications. A filter screens the data to be presented. For instance, the filter specifications on a special report may direct that only those accidents that occurred in locations 100, 151, and 208, in departments A and B, between 1/1/88 and 6/30/88 be reported. The filters for a periodic report would be specified at the time the report was requested, changing with each new report. The report writer will need a provision for handling text that is attached to a variable. Some typical options given in report writers are shown in Figure 5.7.

Standard reports. Standard reports can be made up for departmental or management use. The departmental reports present data for occupational health and safety personnel to review, and cover test results and other technical data. Management reports contain more summary information. Report writers can handle either kind of data presentation.

5.9 DATA REVIEW AND ANALYSIS

Analysis and the Computer. A common assumption is that if there is a problem, the computer can solve it for you. It can, if you tell the computer how to solve it, step by step, and if you have the right data. Statistical analysis of a problem on a computer is well worth doing because it is

OPTIONS FOR REPORT WRITERS

Report Content

Files to be used
Fields to be used
Filters
Report title
Headings
Footers

Processing

Calculations
Sub-totals
Totals
Statistical measures

Page Format

Lines per page
Page numbering
Control breaks
Line spacing
Placement of report title
Placement of column headings
Placement of footers
Top and bottom margins

Print Options

Print size
Font
Italics
Compressed print
Bolding
Underlining

Fig. 5.7 Typical options provided in report writers.

time-saving, but the user is required to understand proper statistical methods and how the computer applies them before expecting the computer to do its job.

Data stored in the occupational health and safety system contain valuable information, but getting the information out takes judicious review and analysis before a report can be written. The computer can be a highly productive tool in this task when the user applies it properly to statistical analysis. The first requirement is to decide what needs to be reviewed and what the objective of the review is. Health and safety professionals should consult a statistician unless they are certain that the data they have selected for analysis can be statistically tested for the objective they have in mind.

Some of the newer statistical packages on the market, called *data interpretation systems*, are written for less informed users. When users do not feel competent to make judgments about their use of statistical measures, they should seek the advice of a statistician.

Levels of Analysis. Perusal of standard reports on test results occasionally indicates an area of interest that needs exploration. Using the query facility, the data can be empirically reviewed and assessed for their potential importance. In succeeding steps, the data can be subjected to alternating review and statistical testing. The levels can be identified and used as follows:

- Level 1
  Review—Select data for review by querying the database and divide the data into groups.
  Analysis—Obtain descriptive statistics for the groups.
- Level 2
  Review—If the data seem promising, place them in separate files for export to more rigorous statistical analysis in SAS, SPSS, or BMDP.
  Analysis—Select appropriate statistical tests for analyzing grouped data.
- Level 3
  Review—Run tests on grouped data and review results.
  Analysis—Do the results appear to apply to the original objective? Is there anything significant about the results that shows the occurrence of the observed problem to be more than mere chance?

An example of data observations that would prompt review and analysis might be the fact, revealed in a standard report, that employees in one plant location have had slightly elevated results on certain blood chemistry tests over the past six months. Following up on these indications, the industrial hygienist may prefer to test that group against another group assigned to similar work tasks in another plant location.

Benefits of Ad Hoc Data Review. The immense benefit of ad hoc review of data is that it gives the occupational health and safety professionals an opportunity to peruse data of special interest. Database queries provide the ideal method for tentatively identifying important information. Once identified, the information can be reviewed and tested before reaching any conclusions.

Comments on Level 1 Analysis. At this stage, statistics should be available in the integrated occupational health and safety software, including simple statistical measures like sample size, range of values, arithmetic mean, median, standard deviation, and variance. Programmers should be given explicit directions by someone knowledgeable in the statistics used in order to code the procedures correctly. Industrial hygienists may require standard error and 95% confidence limits for their work.

The DBMS should cover these simple statistics and should include frequency distributions and simple scatter plots. Whether the statistics available to the user will come from the application package or the DBMS is a matter for discussion with the developer during design. A fundamental task in data analysis is to determine the charcteristics of the population observed. In addition to measures of central tendency, it is helpful to know the distribution and variability of the population.

Since data can be exported from the occupational health and safety system to a statistical package, expansion of the statistical analysis at this level beyond basic measures is not necessary. In the statistical software, a histogram will show the shape of the data distribution, a scatter plot will illustrate any outliers, and frequency tables will give frequencies for categories within the fields. Skewness and kurtosis of the distribution curve will add to the evaluation.

Comments on Level 3 Analysis. At the final analysis level, the statistics used will depend on the information that has to be evaluated and the objective of the analysis. At this level, one may need the advice of a statistician. Among the statistical approaches that should be available to the occupational health and safety users are regression analysis, analysis of variance and covariance, and cluster analysis. In evaluating assessments for ranked exposure groups, industrial hygienists will employ nonparametric statistics, including many of the more common and less complex measures just mentioned.

Data Analysis and Interpretation. Data analysis begins when data are manipulated. Where test results are transformed, such as correcting audiogram results for presbycusis, data analysis begins, the purpose being that a somewhat different view will rule out variability like the effects of aging. Industrial hygienists use geometric means across sampling results because the geometric mean is more representative of their sampling distributions, but for geometric means the values are transformed to logarithmic values first.

Data interpretation follows data analysis. The validity of the interpretation is directly proportional to the validity of the analysis, which is why it is important to select the right data and the right statistics to match the objective of the analysis. "Right" is a term relative to the viewpoint of the user. Interpretation of data is subjective, although we like to consider ourselves objective by presenting facts that are statistically tested and upon which we base our conclusions. Some interpretations are so nearly alike that they are accepted as gospel, but most are accompanied by a measure of doubt, and these subjects become controversial. There is no shame in being controversial so long as the data analysis and interpretation can be reasonably substantiated.

Statistics assumes a mathematically explainable pattern, but in the real world this idealized pattern is rarely, if ever, observed. The patterns are models that we compare to real life situations. It is like having a group of people describe the appearance of American women compared to that of a top flight New York model. Some would compare in terms of physical measurements; others would describe physical features of hair and eye color and skin texture; still others would combine both measurements and features. There will always be mavericks who contend that none of this is a reasonable standard for comparison because brains and personality were not accounted for. All of which is analogous to the data analysis and interpretation in occupational health and safety. The variability in viewpoint is an inescapable hazard in constructing a statistical data model and interpretation.

Statistics Packages. There are literally hundreds of statistics software packages on the market for all sizes of computers and all kinds of applications, but only a small number of these lend themselves well to occupational health and safety. The three that have been commonly used on mainframes and minicomputers are now available for microcomputers. They are modularized to keep the price down, the buyer purchasing the basic package and selecting additional modules as needed. All three support import from, and export to, files in other software through ASCII transfer.

BMDP (Biomedical Computer Programs) was developed under the National Institutes of Health (NIH) Special Research Resources Grant RR-3 at the Health Sciences Computing Facility, University of California at Los Angeles. For two decades this has been a standby in the biomedical fields, but there are ample tests to suit the needs of industrial hygiene and safety as well. Because of its age and mainframe origin it is not very user friendly, even in the 1988 microcomputer version. The reason is that the user has to learn the procedural language needed to give the program its commands. The original BMD programs were written in

1964 only for their computational value, and contained no user-friendly features, making them hard to learn and hard to read. The BMD-P series was distributed in 1969 for easier use and improved statistical methods. What this package lacks in user friendliness it makes up for in thoroughness and fully tested accuracy.

Statistical Analysis System (SAS) is widely used and operates with a language that has to be learned. The documentation is complete and extensive. It is an interactive system with menus. This package takes a lot of memory, so microcomputer users may find it unsuitable unless they have upgraded their system's memory to the megabyte range.

Another venerable package is SPSS (Statistical Package for the Social Sciences). It is interactive, with menu and help screens, and also operates by command language. The microcomputer version is more flexible than the mainframe and minicomputer versions.

All three packages contain programs for graphic plots. These statistical packages were developed for comprehensive statistical analysis, and they provide many valuable calculations. Because they operate by a command language which has to be learned, and because of their comprehensive coverage, they are best used by the serious professional who is willing to work at learning how to use them. In occupational health and safety, the need for statistical analysis is always present. The primary concern is that the user know what statistical measures are suitable to a problem and how to apply them. For a good review of many of the available microcomputer statistical packages, the reader should refer to popular microcomputer magazines.

Reporting the Data and Interpretation. Having drawn some conclusions from data interpretation, the final task is to report it. This can be as informal as testing an idea and as formal as a written report to management about your view of conditions. In testing an idea, continued curiosity can lead to new ideas and further tests. One of the most effective features of the occupational health and safety system is the ability to look reiteratively at data in the system on an ad hoc basis and test them.

Written reports can be as formal as the situation demands. Scientific presentation requires a rigorous explanation of the sampling and statistical analysis, and a formal writing format. Reporting to management requires that the conditions and interpretation be explored in the body of the document, with the analytical methods placed in the appendix. The occupational health and safety system furnishes the data and subsets user-selected data for export to the statistical package, where statistical analysis occurs. Readers of reports can do their own interpreting when data and methods are described.

5.10 PROFILES

Types of Profiles. A profile is a panoramic view that attempts to portray the outline of an individual or condition in relation to one parameter. The parameter can be a person's physical condition or the hazardous exposure outline of a location. By reviewing the profile, an informed assessment can be made about the person's health or the safety of a location. Profiles are very useful reports in occupational health and safety, although they are more typical of the medical profession than of safety or industrial hygiene. Physicians use individual profiles of a patient that include medical test results, vitals data, medical history data, and family history data. Other profiles in health and safety could be exposure profiles on individuals or locations, or an entire plant could be profiled for illness, absenteeism, incidents, or exposures.

Profiles can be done on a regular or special basis. They are summaries that illustrate an overall view of a person or a condition. This broad set of data can be quantitative, qualitative, or both. The profiles can include only one set of test results or succeeding sets of test results over a period of time.

Purpose of Profiles. Profile evaluation is a basis for action if the condition of the employee or location appears to be threatened. Additional medical or industrial hygiene surveillance may be needed to prevent further damage. The profile can also indicate that everything is under control, so it can feed back positive as well as negative information. A series of profiles made on the same individual or location can reveal trends over time that necessitate action.

A Company Profile. One type of company profile relates the number of absentee days to absentee locations. When the locations are geograpic areas, the profile indicates where absenteeism is occurring but does not identify the cause. Its effectiveness is in bringing to attention those locations that need to be examined for possibly unrecognized problems. Hopefully, the problems will prove to have off-the-job rather than on-the-job causes.

Most companies attempt to track absenteeism for production purposes. However, the tracking suggested here is a proactive effort to pinpoint incipient hazardous exposure.

An Individual Profile. Physicians review patient conditions by profiles, or they present the patient with a profile to support an opinion. Some of these profiles are poorly formatted, which detracts from the clarity of the report. Information is scattered all over the page, ungrouped, and in random order.

There are two ways to present profiles: a vertical list or a matrix consisting of rows and columns. In a vertical list, there should be only one set of test results and information—all the data pertaining to a recent physical examination, for instance.

In a matrix, rows would consist of test results and columns would cover time. The time can be designated as intervals of time, numbered examinations, or simply dates of examinations. The matrix profile is more difficult to program, and it has some disadvantages. Dates on many tests usually vary, so the column headings may not be specific unless they cover periods of time. The advantage of this type of profile, however, is that the reader readily recognizes upon inspection any changes or trends in the patient's health (see Figures 5.10a and 5.10b).

Vertical lists of data should be grouped for easy reading, and spacers and headings can visibly separate groups. Profiles usually contain physical examination findings, results from tests like electrocardiograms, X-rays, pulmonary function tests, laboratory tests, and personal and family history data, all of which lend themselves to easy grouping. When formatting profiles is easy, why they are presented in a hodgepodge manner is a mystery!

A Multipart Profile. In an integrated occupational health and safety system, it is possible to generate a multipart profile on an employee. Although this is one report, it actually consists of several profiles. Profile data come from occupational medicine, industrial hygiene, and safety. The medical report consists of the history, test, examination, and clinic visit data; the industrial hygiene report includes all exposure data from both area and personal monitoring; and the safety data include the investigation and incident cause data on any accidents or injuries. This comprehensive profile is too extensive to do on every employee on a routine basis, but it serves a purpose in compensation cases or situations where a complete occupational health assessment is needed for an employee.

5.11 GOVERNMENT-MANDATED REPORTS

Designing Mandated Reports. State and federal government reports can be designed into the computer system provided that they are formatted and read like the government-prescribed report. A facsimile must contain the exact wording and arrangement to be acceptable. Horizontal and vertical lines can be replicated on the computer form.

MEDICAL PROFILE REPORT

```
NAME: Jane B. Sanderson               REPORT DATE: 06/07/1988
EMPLOYEE ID: 33333                    SEX: F
SOC.SEC.NO.: 333-33-3333              AGE: 29

JOB CODE: 24680                       JOB DESCRIPTION: Maintenance
EMP. STATUS: FT Full time             PROCESS: 14 R&D
SHIFT: 1 Day                          DEPARTMENT: 432 Chem. Lab.
                                          LOCATION: Bldg. C
----------------------------------------------------------------------
PROBLEM LIST:
  PL 444  Hypercholesterolemia
  PL 332  Hypertension
----------------------------------------------------------------------
CLINIC VISITS
DATES:               03/15/1987       05/09/1987       09/28/1987

Reasons:             01 Complaint     02 Sched.Test    08 Supv. Ref.

Weight (lbs.):       179              184              185
Height (ins.):                        68

Blood Pressure:      165/104          172/104          156/109
Pulse:               85               86               92
Temperature:         98.0             98.6             102.4

Complaint:           064 Headache                      002 Fever
Disposition:         10  Medication                    01  Sent Home

Allergies:

Medications:
 Mevacor
 Aldomet
```

```
MEDICAL PROFILE REPORT  Jane B. Sanderson       06/07/1988       PAGE 2

TESTS AND EXAMS:
DATES:                                   05/09/1987

Audiogram:
 Right Ear STS
 Left Ear STS
 Binaural Imp. %                         0

Pulmonary Function:
 (% Predicted)
 FEV(1)/FVC                              76

Electrocardiogram:
 Resting
 Interpretation                          121 Tachycardia
```

Fig. 5.10a An illustration of a medical profile report on an employee.

```
MEDICAL PROFILE REPORT  Jane B. Sanderson        06/07/1988        PAGE 3

DATES (cont.)                                  05/09/1987

Blood Chemistry:
 BUN                                           16
 Cholesterol                                   247

Urinalysis:
 Appearance                                    Clear
 Color                                         Lt. yellow
 Acetone                                       0
```

```
MEDICAL PROFILE REPORT  Jane B. Sanderson        06/07/1988        PAGE 4

DATES (Cont.):                                 05/09/1987

Hematology:
 WBC                                           9.6
 RBC                                           5.2
 Platelets                                     3.5
```

```
MEDICAL PROFILE REPORT  Jane B. Sanderson        06/07/1988        PAGE 5

EXPOSURE RECORD:
 Exposure date       02/24/1985            3/09/1986
 Agent               Chloroform            Acetaldehyde
 Estimated exposure  252                   520
 Measured exposure
 Units               ppm                   mg/m3
 Reference standard  NIOSH Ceil 220 ppm    OSHA 360 mg/m3
 % Exceedance

PPE:
 Date assigned       01/05/1985            4/15/86
 PPE assigned        Gloves, Neoprene      Disp. Coveralls
 Type equipment      Hand protection       Clothing protection

PAST WORK HISTORY:
 Start date
 End date
 Job Code
 Description
 Process
 Department
 Location

Work Restrictions:
 Restriction Code    222
 Description         Limited stair climbing
 Reason              Cardiac enlargement
 Begin date          05/09/1987
 End date
```

Fig. 5.10b An illustration of a medical profile report on an employee (cont.).

The time it takes to design and program the report into the computer is worthwhile, since a computer report to retrieve the information has to be designed anyway. The only additional effort is in formatting, and once formatted, the report can be produced as many times as needed.

OSHA 200. The OSHA 200 log can be replicated by programmers despite its width, numerous columns, and headings. The narrow columns for illness (or injury) classes and lost time have matching headings that are vertically stacked and indented to match the columns. None of the extensive explanatory information at the tops of the columns has to be repeated, but the headings must be identical. Care must be taken to number the columns exactly as they are numbered on the prescribed form because all of these data are lifted by the BLS for their annual records. This is why the format must be scrupulously preserved.

On a wide printer, no squeezing will be necessary to accommodate all the columns required, but 8½ × 11-inch pages present problems, since printers are not oriented to shift the printing ninety degrees. The solution is to use compressed print and to print in the normal vertical position. A photocopy can be enlarged and printed on the long dimension of the paper.

At the bottom of the computer form, there must be a place for the signature and certification. A simple record should be retained in the computer database, giving the date of preparation and when the report was forwarded, along with the name of the person who signed and certified it.

Totals must appear according to the required format at the bottom of each page, with cumulative totals also shown by page. Pages must be numbered.

A constant problem with this report is keeping track of lost workdays and updating the report to reflect changes. Having this report in a computerized system overcomes the headache. As the employee record is updated, the OSHA log can be rerun. Although the log lists 18 employees per page, the computer report can carry whatever number fits the page.

OSHA 101. The form for the OSHA 101 does not have to be used as offered. If the First Report of Injury/Illness covers the information required, that can be submitted instead. The format of the computerized report does not have to emulate the government form, and the extensive directions can be omitted.

Rather than prepare a 101 report for each ill or injured employee, it makes more sense to format the First Report record so that it can be output directly to the printer as a 101 report.

Hazard Communication Reports. MSDSs should come off the printer exactly as they appear on the computer screen. Reformatting, filtering, or processing any of the information should be avoided, since MSDS data are supposed to be complete. Where data for a prompt do not exist, there must be an explanatory remark like "None," "Not Applicable," or Not Recorded." All prompts must carry some kind of response.

MSDSs have to be available for three types of people: employees, customers, and the community. Menu selection of one of these should be programmed in order to prompt the user to give the name and address of the recipient and the name of the agent, and then record the transaction with that date in a file of persons to whom the MSDS was sent.

Notifications to employees about medical tests and industrial hygiene monitoring results are actually highly structured reports. As many different letters as are needed can be written, the forms stored in the computer, and a menu programmed to cover them. Selection of a menu item would be followed by a prompt indicating which employees and dates are to be covered. Appropriate programming would extract the required data from the databank, insert them in the report, date it, and record the transaction in a file for the purpose.

EPA Tier One, Tier Two, and Form R Reports. These reports may be tedious to program, but again, the effort is worthwhile. The main success with computerizing these reports will come from good organization of the database. If agents carry purchase, storage, and usage data with specific locations, these reports should be reasonably straightforward. Data describing the agent such as Chemical Abstract Society (CAS) # , chemical name, hazard classification, and state, should come directly from the Agent Identification record. At the end of the document, a certification statement must be added with a signature. A file needs to be kept on the date of report preparation, the signer, and when the report was sent. Since the information about sending OSHA and EPA reports is alike, it makes sense to have only one file for these records of reports sent, with the type of report identified in the record.

Other Government Forms. Manifests for transport of hazardous materials should be designed according to state and federal government formats. The data should come directly from files attached to screens carrying the data. Most of the data are so specific that they must be entered in screens designed for that purpose.

Because there are so many different permits, both state and federal, it is not worthwhile to design a form for each type. Instead, it will suffice to keep a permits record and include renewal reminders in the tickler file.

5.12 CONTINUING THE DEVELOPER-USER RELATIONSHIP

The Jointness of Development. Throughout this book, beginning with the Preface, emphasis has been placed on developing the occupational health and safety system through the joint efforts of people with different agendas. Chapters 1 and 2 dealt with the joint needs of specialists in health and safety, and with the joint needs of company management and the health and safety specialists. Chapter 3 emphasized what the developer needs to contribute while working closely with the users. Chapter 4 placed the emphasis on the users' contribution to development. Chapter 5 continued the joint effort of users and developers, but with lines that were not always clear-cut for the contributions of either, and with considerable overlap of effort.

What follows system development remains important to both developer and users. If this is a commercial system, future sales rely on the successful use of every system sold. The occupational health and safety field is a small community that communicates well within its professions. News of successes and failures travels quickly.

If the system is developed in-house by information services, successful use of the system will mean that minimal assistance is required by the user departments. It is also a political plus within management circles that both developer and user can point to a successul project. The developer's involvement will diminish over time, but the user is expected to obtain maximum benefit from this large expenditure.

A Continuing Relationship. The real truth of the user-developer relationship is that it is unlikely to end. Just as users will operate their system in future years, developers will continue to be involved as needed for system maintenance and repair. If the occupational health and safety system is commercially developed, the developer will undoubtedly come out with new releases and updates that enhance the original system and eliminate elusive bugs. Most contracts require vendors to keep user systems updated. For the in-house developer, users invariably find new things they would like to include in the system, and that means more development.

Like all appliances, hardware develops small bugs, and users have to call for repairs from their in-house experts or vendors. Not being mechanical, software does not require repair because of breakdown. Software repair occurs during the early months of use when something different is tried and a previously undetected bug is uncovered. There is nothing unusual about this occurrence, and the developer can fix it without detriment to the growing database, despite the patch to programmed code.

Software maintenance is a long-term responsibility. As users explore the capabilities of the system, they are bound to get new ideas or want to add new programs. The one certainty about software is that the user always finds something else that should be added to it. Software developers seldom terminate the project because, by the time they release a package, they already have new ideas to add, or ways to make the present software more efficient or more usable. Software maintenance is the process of adding to the existing package.

The critical requirement for successful repair and maintenance is good programming documentation from the start of the project. This allows any new individuals who work on the code to approach the assignment knowledgeably, with as little disruption to existing code as possible.

Users should be aware that repeated additions to the software can eventually be detrimental. As Figure 5.12 shows, there comes a time when additions fail to enhance and begin to cause unresolvable problems. This happens to all software, no matter how careful the programmers are. Commercial software developers have this problem after two or three revisions, each with numerous enhancements, which is why a developer with a new package can overtake the sales of the old and trusted package. The original package is just too overloaded with programming patches and additions. Software has a life cycle, and there comes a time when technology outdistances the original package and the developer must begin anew. Hopefully, this can be done by writing new code that will accurately accept the transfer of the old database to the new one.

For code developed with 4GL tools like screen painters and report writers, numerous modifications and additions are easily accommodated because they generate code without programmer intervention.

5.13 LOOKING TOWARD THE FUTURE

Sharing Resources. Computer technology is rapidly moving in the direction of shared hardware and software. From the days of large mainframe machines that everyone shared, we moved into the microcomputer era, where everyone clamored for his own desktop machine. This was a natural reaction to having been forced into the mainframe mold, where the software on hand and one or two database managers had to serve every company purpose and information services dictated the choices. Independence came with the desktop microcomputer but had the disadvantage of eliminating shared data.

The increased power of the microcomputer is making it possible to

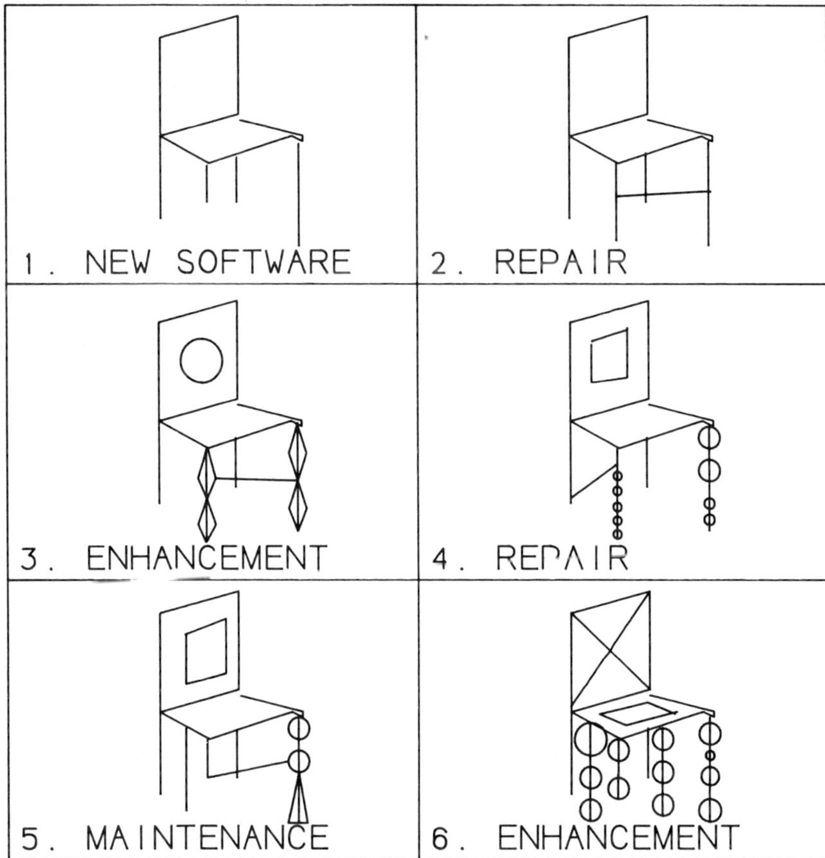

Fig. 5.12 The effects of continued repair, maintenance, and enhancement on a software system.

emulate the mainframe on the desktop and to share commonly used data. A multiuser system of 5 to 15 people makes this possible.

The increased power of microcomputers is a function of computer power, speed, and expandable memory. New operating systems will be multitasking, which means that more than one software package will be usable at any one time and interfacing will be simplified.

Improved software and hardware for linking machines as shared systems are increasingly drawing users together, which is beneficial and cost-effective. It costs less to purchase run-time software for several users than to purchase individual packages.

Sharing Occupational Health and Safety Resources. Contradictorily, the field is called *occupational safety and health* (all in one breath), but in practice it is three distinct fields: industrial medicine, industrial hygiene, and industrial safety. The objective is to protect employee health and the environment, which requires the services of all three types of professionals, each making a contribution. Somehow, the common objective gets lost in practice and each discipline tends to follow a separate path, with only occasional convergence of interests.

The intent of an integrated occupational health and safety system is to provide shared information resources. To accomplish this, attention must be focused on a common goal. At present, computer hardware and software are purchased because of the initiative of one or two disciplines. Rarely are all three types of professionals equally anxious to buy or develop a system. The end result is that an integrated system loses cost-effectiveness because it is not used in the manner for which it was designed.

When each professional finds small, inexpensive packages to automate a task, he or she buys a microcomputer for that one purpose. It would be much more cost-effective if four or five different tasks were served by one comprehensive software package.

Integrated Software. Ideally, one purchase of application software should cover all of one's recordkeeping tasks. Earlier discussions showed some limitations to this ideal because of the different programming approaches for graphics, statistics, modeling, and word processing. Some day the limitations will be overcome, but for the present, we can make more effective use of what is available in integrated form.

This means that the hygienist who is attached to Lotus, the safety manager who is attached to an incident investigation package, and the physician or nurse who is attached to a hearing conservation program should emerge from isolation and pursue occupational health and safety as a tripartite team. True integration in occupational health and safety is yet to be realized throughout industry. A good beginning is the team development of an integrated data management system. The opportunity to share hardware and software will support real integration of the health and safety disciplines into a unified occupational health and safety systems.

Development Resources. With improved 4GL tools for design, development, and prototyping, the costs of producing an integrated occupational health and safety system are being reduced. This does not imply that the cost of such a large system is likely to be low. Large systems will always be relatively expensive. What reduced cost will create is a better chance that commercial vendors will produce occupational health and safety software, and the product's cost can be reduced enough to make it marketable.

Up to the present, the integrated occupational health and safety system has been so expensive to produce that the cost deterred all but the Fortune 500 companies, which developed their own in-house systems. Small businesses were unable to purchase commercial packages because the price made the software totally unobtainable, and they were without in-house resources to undertake development. Yet, small businesses need this software. Lacking sufficient in-house development expertise or the money to buy a commercial system, many professionals have settled for single-purpose dedicated packages on a microcomputer, satisfying themselves that they were automating their tasks. And indeed, individual tasks benefitted, but not the much larger responsibility of automating occupational health and safety data management.

Improved Output. The outstanding feature that sells any software system is highly useful output. The emphasis should be on quality, not quantity, output. Quality is judged by the system's usefulness to a company, its professionals, and its management, not by the developer's opinion. Improved quality of output will come with efforts to sharpen report skills. More emphasis must be placed on using computers to process and output data in meaningful reports. Report generators and query languages make this possible.

Report generators make it relatively inexpensive to generate a number of standard reports, but government reports present problems. These reports must be programmed because they are reproduced, not produced. It would help if government agencies designed their reports with computer users in mind and made their reports compatible with users' computer resources. The government designs printed forms that have no easy reproducibility in a computer. With many employees and agents, a company is sorely pressed to fit itself into the government report mold.

Output will continue to improve with greater use of query languages built into database managers. SQL appears destined for stardom, but suffers at the moment from too many variations. This problem is being resolved through standards being devised by ANSI. SQL is now being used with microcomputers, and its use will increase with improved operating and database management software. OS/2 Extended edition will have an SQL database built into its operating system. The key to setting up meaningful reports and subsetting for analysis is to query the database first.

Leading the Way to the Future. The need for good occupational health and safety management carries with it the imperative need for improved recordkeeping. Thus far, government has taken the lead to set standards for records to be kept and reports to be made. Why should government have to force industry to maintain records and mandate that they be shared by employees, unions, and the public? Why is it that professional organizations do not assume a leadership role in promoting good record-keeping by devising guidelines for their members to follow? There should be guidelines for data that need to be kept, guidelines for evaluating them, and guidelines for putting them to good use, followed by professional development seminars to train users. The only development seminars on recordkeeping are held in reaction to government mandates for records. This is reactive, not proactive. A cursory examination of professional occupational health and safety journals shows pages and pages of seminars and workshops on accident investigations to zero mechanical states, absorption coefficients to zero concentration, and abdominal injuries to zoonoses. No workshops exist on efficiently managing data from A to Z, which is what follows all the other As to Zs.

Lacking expertise with recordkeeping, users are handicapped when vendors offer them computer systems to automate their tasks. The result is that there are many partial efforts to deal with a very large, important business. No one in his right mind would purchase only an accounts receivable or accounts payable package. The user wants the whole range of software, from payroll to general ledger and journal. The automation of occupational safety and health is a critical business. It deserves to be upgraded to the level of business accounting, and the users themselves must increase their importance to the company through both individual effort and professional organizations. They can lead through their combined talent, power, and resources, as has been ably demonstrated by their efforts to train and certify professionals in occupational medicine, industrial hygiene, and safety management.

Instead of complaining that the government is costing industry millions of dollars in recordkeeping and reports, occupational health and safety professionals should be part of the solution rather than part of the problem. They should be proactive instead of reactive, and should improve recordkeeping and reporting for occupational health and safety. There should not be just medical records, industrial hygiene records, or safety records, but instead an integrated database of all this information to evaluate the health and environment of employees in the workplace and the community. This can be done without loss of confidentiality.

5.14 SUMMARY

After the developer completes and tests the new occupational health and safety computer system, hardware must be installed, the new software installed, and systems and user personnel trained to run it. Old data may be entered, and there will be a period of simultaneous operation of the old and new recordkeeping methods, called *parallel processing,* before complete confidence is placed in the new system. It takes time to evaluate the system properly, and acceptance should not occur before a full evaluation has been made.

Training and documentation are designed differently for systems personnel and users. The more user friendly the system, the less user training will be required. Planning and conducting training depend on cooperation between users and developers. The training may take place on-site or at the developer's shop. Training aids, training manuals, and a skilled trainer are important. Dictionary codes should be in place before training begins, and users should schedule their work to permit giving their undivided attention during training.

Documentation is a permanent reference guide to successful use of the system. An operations manual serves the systems personnel. For systems management there is a manual describing the system's architecture, functions, and capabilities. A reference manual contains fields, files, and code descriptions. Users need a training manual and a manual describing in clear, concise terms how to use the system. Screen painters, report writers, and programming code each have their own documentation. Help screens provide on-line documentation for user convenience that can be very helpful if carefully designed for user friendliness. All documentation should be evaluated.

Data are processed in the system and, upon retrieval, become information. Retrieved information is interpreted in the human mind and becomes knowledge, which directs our actions. Users must decide whether the computer is only a data repository or an information processor. Professionals will want it to serve the latter purpose. Occupational health and safety information is communicated to management through reports.

Reports need to be carefully designed to convey information reliably. Of the many types of reports, some are for ad hoc use, others for periodic use, special purposes, and to fulfill government requirements. Ad hoc reports are readily available through query language. Queries are also valuable for reviewing data prior to subsetting them for rigorous analysis.

Review and analysis is an interactive process that gives the occupational health and safety professional an opportunity to determine which data describe an observed trend in the employee population. Data can be transformed in the occupational health and safety computer system, subset, and exported to statistical analysis software such as BMDP, SAS, and SPSS. It is important to select statistical tests that match the purpose of the analysis and the data used.

Profiles portray essential data about an employee's health, hazardous exposures, or company conditions, either at a particular time or over specified periods of time.

Government-mandated reports can be computerized if they are a facsimile of the prescribed form.

The developer-user relationship will continue for the life of the software. Once the bugs in the original system are repaired, the user will want to make changes and additions as time passes. This maintenance work should be requested sparingly in order not to disrupt working programming code. This problem will not be as acute when a screen painter and report writer are used because they generate code automatically.

The trend toward shared hardware and software is favorable for integration of occupational medicine, industrial hygiene, and safety data into an occupational health and safety computer system. Small businesses are especially in need of integrated recordkeeping systems for health and safety. Quality computer reports will come with increased skill. Report generators and query languages are the tools that will make this possible.

Occupational health and safety professionals should take the lead in providing guidelines for their recordkeeping through their respective organizations and seminars. By promoting improved data management, they will be helping themselves and will reduce governmental control.

Appendix 1

THE SPECIAL CASE OF THE SAFETY MANAGER IN A SMALL BUSINESS

Special mention must be made of the lone safety manager who oversees all occupational health and safety within the small plant. In some cases, a nurse or nursing staff takes care of employee illness and injury under the supervision of the safety manager. In other cases, health care is contracted out and the safety manager is the only health and safety professional on the job. This safety manager is the integrating element for health, industrial hygiene, and safety; he operates on a nuts-and-bolts budget, with a microcomputer on the desk. In the eyes of management there is no justification for an expensive integrated occupational health and safety computer system, nor are there usable small packages on the market that will cover all of the manager's needs.

This is the individual who relies on dBASE III Plus and Lotus 1-2-3. dBASE has some programmable language, in addition to being able to handle files, which allows the safety manager to file data and retrieve them. Since the calculations offered in dBASE are limited, files created in dBASE can be transmitted to LOTUS for numerical transformation and manipulation.

After practice, the manager may gain the confidence to use dBASE commands, and small programs are written that accomplish basic data management. LOTUS has macro capabilities, and after practice the user can instruct the machine to accept macros. A macro is a user-specified set of characters that performs a given procedure. A procedure may be one to several commands that perform a function. Set the macro for the procedure, and thereafter at macro entry, the procedure is automatically kicked off. This is a shorthand message from the user to the computer. LOTUS permits generous use of macros.

In all probability, the safety manager was convinced by software ads, popular computer magazines, a computer enthusiast, or a salesperson that dBASE or a similar package is a great database tool for only a few hundred dollars. Buy a microcomputer to go with it, and you are in business! This is true only if the buyer knows the limitations of both the software and the hardware, and if limited growth potential is expected. The limited nature of dBASE and LOTUS eventually becomes apparent, and the safety manager may not understand why.

The safety manager's enthusiasm about working his own computer system is the driving force that overshadows the strengths and weaknesses of databases and of the software that controls them. In the past, one of the marked differences between software for microcomputers versus larger machines was the ability to handle databases. Microcomputers now have the large memory capacity necessary for DBMSs.

The safety manager should understand the differences between file and database management, and between database management and a DBMS. File management is done by programming every step required to process records and files in the computer. This means that a programmer has to tell the computer how to process the data, store them after processing, and retrieve them. File management methods on the mainframe were an early development, and demanded a lot of programming time and expertise. In file management the files can be created; data can be added to the files, modified, and deleted; and data can be retrieved. However, there is no control over the files beyond what the programmer specifically writes into the software.

Database management evolved from file management, with the added feature of specific database languages to address functions like data description and data manipulation, as well as data storage and retrieval. These functions are transparent to the user, and the user is free to create files, add to them, and delete them, because all the necessary housekeeping has been programmed behind the screen. This database management technology was a vast improvement over file management because it allowed the user to create and use files without direct employment of programmers. This is the advantage of software like dBASE; it is user oriented. The command language that is part of managers like dBASE gives the user additional freedom to build screens for data management. The database is usable without the refinement of screen forms. Although file housekeeping is done by the database manager software, the user has to program everything else that pertains to screen display.

Using the commands that work with dBASE and similar software, a safety manager can turn programmer, but the product is likely to be limited. It takes a truly creative programmer to maximize the use of languages. I have seen a great deal of dBASE programming, including programming by experts with a lot of experience, but I have seen only one person who was so gifted at programming that dBASE could be made to carry out almost any task. Even with this extraordinary talent, writing a creative program presents problems if this person leaves the job and future maintenance and repair are required. Reliable programming is more than just writing the original program; it has to be able to undergo changes in the future. Room for growth is essential to software.

The next higher level of database management came with DBMSs, which improved on database management by handling not only the database, but the screen display as well. These remarkable software packages also provide data dictionaries, data security, and data integrity checking. This type of software is much more expensive than a database manager. New versions are multiuser oriented and have query language facilities, making them a far wiser investment, dollar for dollar, than a database manager. The disadvantage is that they are more difficult to learn than software programs like dBASE. They are now available for microcomputers, but they require large memory capacity on hard disks. 4GL screen development provides the front end for the database, and report writers further enhance application software development. They provide good menu creation facilities, whereas menus for dBASE are written by the programmer.

Another advantage of DBMSs is their ability to create indexes. dBASE will create an index over a file when specifically directed to do so. However, when the file is used, the index has to be updated, and it is the user's responsibility to request the update. This becomes a work practice issue that requires the user to take time each day to update indexes for the files that have been worked on. When the file reaches a large size, reindexing can take time. Reindexing frequently during the day is time-consuming; leaving the system turned on over the lunch hour or overnight may be the answer. A DBMS provides improved means for updating the indexes attached to files, relieving the user of index-tending burdens. A short program can be written for the user's convenience to set off the indexing procedure, but the user may need to state specifically which indexes should be redone.

The latest tools for microcomputer software are 4GL screen painters and report writers that have built-in database capabilities; they vary in price according to their level of sophistication. For the user who wishes to develop a small occupational health and safety system application, this is probably an improvement over data managers like dBASE, and growth can be achieved. However, while screen painters with attached databases can be purchased for only a few hundred dollars, they do not include some of the special flexibility of occupational health and safety software. A real occupational health and safety screen painter is an expensive item for a small business safety manager, and may require more time for development than is available.

However small the company, a safety manager needs a health and safety computer system. Information needs are just as acute for this person as for the corporate safety director, the chief difference being the volume of data handled. The same data are needed by all occupational health and safety professionals. Without doubt, there should be such a facility on every safety manager's desk. The basic problem for the small company is money, a problem that a book on occupational health and safety software cannot solve. The solution lies within the company—its commitment to helping the safety manager and its ability to pay the price.

Beset by the need for a database system and hampered by limited funds, many of these safety managers have opted for putting files on dBASE and similar database managers. They have found that the flexibility of the application is limited, and each time they have tried to expand it, problems were encountered. They should understand that large companies have been equally stymied in the past. Older occupational health and safety systems were programmed at great expense and worked fine for a while, but growth created trouble. The problem has not been with overly ambitious users or programmers unequal to the task, but with the state of computer technology. We have grown through file management, database management, and DBMSs. We have grown through dependence on mainframes for databases, then minicomputers, and now we have microcomputers that can handle DBMSs and front-end machines that can handle application development. For occupational health and safety, there are now screen and report painters specifically designed for applications software that will work over excellent DBMSs on new, large capacity microcomputers. The evolutionary process has caught up with industrial health and safety needs, with some exceptions. There remains a need to incorporate statistical and graphics capabilities into the software, as well as the capacity for cost management. These will come in good time.

Hundreds of thousands of dollars have been invested in the present state-of-the-art software, which is aimed at a very special market. Realistically, it is difficult to imagine that the price of an occupational health and safety computer system will ever reach the level of popular software and hardware. However, in companies of every size, safety managers, nurses and physicians, and industrial hygienists will be harder pressed in the future to automate because their information demands will continue to increase. The demand for occupational health and safety software exists, and in time prices will be reduced. Hopefully, this will work to the benefit of the safety manager in the small business.

To cope with the present pressures, safety managers in small businesses should recognize the constraints of small user-programmed systems and limited budgets, with no need for apology. Professional organizations must play a large role in providing reliable information on how to computerize. An important step is to seek specialized help from software specialists. Journals frequently have papers describing the way someone programmed or computerized one facet of the health and safety job. Papers presented at meetings do the same thing. Nevertheless, these are fragmented computer efforts rather than software for an application, and they reveal a lack of insight into computer technology. Professional development seminars show the safety manager how to run a computer, how to use dBASE, and how to use LOTUS. What they fail to realize is that software is a field in itself, a field that the user has neither the time nor the inclination to master, and that applications software development should be done by specialists for a maximum dollar return.

Appendix 2

A SHORT COURSE IN CODING COMPUTER DATA

REASONS FOR CODING

Coding is an important recordkeeping tool that is easy to use, and enables users to retrieve data efficiently and reliably. Retrieval of data from narrative text causes the computer to do a word-by-word search for matching key words. For each query, this is costly in processing time and system overhead. When the computer search of a coded list takes place, there are methods for limiting the number of comparisons that have to be made before a match is found, which saves processing time.

Most recordkeeping produces unorganized information in as many informal styles as there are undisciplined recorders. Instead of recording raw facts, the recorders select words that convey personal messages. The reader is left with three choices:

1. To discern the fact in the message;
2. To accept the recorder's version of the message; or
3. To interpret the message within the reader's own frame of reference.

The primary reasons for keeping records are to:

1. Accurately record facts or observations;
2. Examine the facts for their meaning; and
3. Develop a consequent action.

In effect, the sequence is to observe, appraise, and plan accordingly. Accuracy requires that the message conveyed by the recorder to the reader be interpreted according to the original meaning. The action following fact appraisal may be to retain the status quo, to repair it, or to take remedial action.

Coding data makes it possible to deliver messages free of personal suggestions like interpretive words or descriptive adjectives. It enforces brevity and it standardizes observations made by two or more recorders. It also reduces recording time and computer processing time. Coding is a means of communication. It provides quick computer access to facts (data) that can either name or describe something, which may be a person or an object. Descriptions are facts that characterize things.

The practice of coding goes back to the ancients. Aristotle grouped things. Later, Linnaeus provided us with a scientific identification system for plants and animals. With as many as 10 million different kinds of living things, we require a method of identifying what we mean when we communicate and exchange information. Linneaus devised a binomial classification system that names a species with a genus. Using this system, we can describe a plant or an animal in readily identifiable terms, and when further definition is required, we can resort to variety within species. Each variety, species, and genus is detailed in our communications to make certain that the organism is unmistakably identified.

As students, we learn the Dewey decimal system to find library materials, and unless we become librarians, that is the extent of our knowledge about cataloging. We file our business papers by the simplest methods, using the alphabet or numbers. Coding is a form of cataloging requiring a little practice and skill once the user begins to look at the codability of data.

Librarians file books by codes that begin with three-digit numbers that are separated into sections, depending on the subject (history, science, sociology, etc.). Following the three digits is a decimal point and more numbers that indicate the sequence of books within that section. A line of additional code holds a letter for the author's last name. There may be other indicators like the number of the volume in a set, but the systematic arrangement of letters is easy for people to follow as they search the library shelves. This illustrates a simple fact: Large categories can easily be classified by small changes in code.

CODING A DEMOGRAPHIC RECORD

A demographic record presents several worthwhile lessons in coding. Although everyone has a name, some names are not unique. We must eliminate the possibility of duplicate names by the use of unique identifiers. Most companies assign employee ID numbers. This works well as long as the company does not reuse the number after an employee terminates. Because some occupational health and safety records must be kept for 30 years, a company should never reassign numbers.

Our government has presumably tried to make each of us unique through social security numbers, but in a system as vast as social security, the inevitable has occurred: some numbers have been duplicated. Therefore, there is good reason for a company to assign its own ID numbers. Moreover, social security numbers are lengthy and complicated by sets of digits. Most companies have devised simpler identifiers.

Employee numbers have distinct advantages over names. Not only are names not unique, but we change names for various reasons. Women marry, divorce, and occasionally return to their maiden identity, all personal choices that threatens accurate recordkeeping. Men have been known to adopt their wives maiden names, and some married people hyphenate their combined names. While these acts of devotion are admirable, the practical reality is that it complicates recordkeeping. We also complicate recordkeeping by adding Jr., II, III, and IV to honor older generations. Should one of these additions be accidentally omitted, the records can be confused. In plants where father and son work, employee ID numbers are the only solution to the problem.

The employee ID number identifies us as things. Each number is unique, and the time-honored use of clock cards with an ID number is our guarantee that the payroll department will provide us with the right sum of money on payday. Assigning ID numbers is a form of coding that makes certain our individual identity.

Describing people (things) is not nearly as simple as assigning employee ID numbers. Personnel records describe us by our home address, telephone number, sex, race, marital status, age, height, and weight. Unfortunately, no one category provides a unique identity, but in combination they can describe an individual. The problem with recordkeeping is how to enter these descriptive features in a uniform manner for standardized reporting. Addresseses, zip codes, and telephone numbers have no unique value. The descriptors we rely upon in personnel records are sex, race, marital status, and age. Sex, race, and marital status lend themselves to coding.

Why should we code sex, race, and marital status for the computer? Why not enter these just as they are? The answer is simply that people describe things differently, and a computer is effective when there is no ambiguity in a description.

Sex is not an entry problem, there being only two choices. The concern is that the data can be entered as Male, M, or Female, Fem., F. And, if there are checkboxes for Male or Female, will the box be checked or carry an X? There are those who code sex to be 1 or 2, which seems to be carrying coding a bit far, with only two possible choices under any circumstances. The computer can count the number of M entries and the number of F entries just as easily as it can count 1s and 2s, so why is it necessary to avoid the obvious choices of M and F?

Race codes are a nightmare! In the racially varied environment of the

United States, this is the most inadequate form of description. For a variety of political reasons, the government requires us to identify people racially, but it only leads us into coding quicksand from which there can be no satisfactory rescue. In "Can computers cope with human races?", the lead article of the February 1989 issue of *Communications of the ACM,* Les Earnest lucidly describes his often hilarious experiences in trying to categorize his race. This article ought to be recommended reading for all Americans and required reading for computer database designers. Earnest says his forebears came from Eastern Europe and the British Isles, and he has the skin color of an Eastern European together with a Semitic nose. On some government documents he described his race as "Mongrel," which consistently raised official eyebrows and led to concern about his mental health.

Earnest urgently reinforces my own reactions to the problems of describing race in personnel records. Several years ago I was asked to assist in a human relations case that involved a parental dispute with the school system over alleged discrimination. I came away from the experience totally convinced that there is no adequate way to describe people racially unless they are aborigines isolated for all time on an island in the middle of an ocean. Human proclivity toward intermarriage on all continents, especially in North America, has all but obliterated racial relevance.

The father in the mediation case ruefully told me that he had no idea how to answer the last census question about the race of his children. While the father specified his own race as black, he was in fact black and Cherokee Indian. His wife was Italian-born and declared white. What race should he declare his two children to be? (The family's religions were as diverse as the races!)

A check with the Census Bureau today reveals that even the government finally recognizes race as an ambiguous human description. At the time I was involved in the mediation case, there were specific categories of races. In its infinite wisdom, the government has now seen fit to add the category "Other." Nevertheless, if we examine the new set of census categories, we still have problems with classifying the man and his children. The present set of choices reads:

White
Black
American Indian, Eskimo, and Aleut
Asian and Pacific Islander
Other

Should the black(?) man's children be declared white, black, American Indian, or other?

It is obvious that the way we describe our race is a matter of personal preference and has nothing to do with our genetic heritage. In occupational health, there are pulmonary function test records that make use of race. In making calculations for predicted values, blacks are evaluated at 0.9 of the value for whites, and whites include everyone who is not black. Medically speaking, blacks are at risk, but is a person who is a combination of black, white, and American Indian races at risk when genetically no one can tell if the risk continues to exist? Perhaps the physician reviewing this person's records should know all the racial inclusions before committing the pulmonary function test to a permanent record.

When asked about the frequently used category of Hispanic, the Census Bureau declares Hispanic to be among a long list of ethnic origins, not race. This is an interesting development in view of the number of companies that include Hispanic in their race classifications but are forbidden to ask personnel questions about ethnic origin.

Apart from realizing the nonsensical meaning of racial categories, we should consider the problem of how to code race for the computer. Using the census list, how does one code for race? American Indian and Asian preclude the use of letter classifications. It would seem more sensible to use numbers instead of letters to overcome these complications. I know of one company whose race code list begins with A = black and B = white. I feel sorry for the data entry operator who must continually have to recall that B is not black, but white.

Racial classification must also be viewed in terms of the changing attitudes of people. Some black persons are rethinking the use of the color designation to define their race, inasmuch as we now tend to describe race differently. Whites are now caucasians, and some blacks are suggesting a preference for African-American. There was a time when American Indians were described as the red race and Asians were described as the yellow race. Color seems to be disappearing from the racial scene. This does not affect the divisions as such, but it should give us reason to foresee possible political and cultural pressures that may alter racial classification in the future. With the continually changing perceptions of our cultures, following the changes with computer codes must inevitably negate the use of racial or marital status as conditions for classifying groups of people, which is a point that is bound to upset the persistent habits of epidemiologists.

Marital status has legal reasons for being a part of our personnel records, and occasionally some medical reasons as well. However, in our modern society, the usual classifications leave a lot to be desired. What is accomplished by declaring that you are single, married, or divorced when people live together without benefit or either marriage or divorce? We might better keep data on whether there is a legal beneficiary or a legally responsible person. There are medical conditions in which a physician may want to know the marital status of an individual, but this can be determined through medical questioning rather than a demographic file. If we must continue to keep records on marital status, our methods for classifying this information should be brought into line with reality. Old classifications die hard, and we should adopt a more critical attitude about the data we keep.

Socrates is quoted as having said that the unexamined life is not worth living. How very true! We could paraphrase this philosopher by saying that the unexamined record is not worth keeping. Some valuable lessons we can learn about coding from the demographic record are that we should:

1. Choose data that can be meaningfully stated on records.
2. Avoid arbitrary coding.
3. Keep codes simple, avoiding unnecessary revision, if possible.
4. Maintain flexibility in our codes for revision and expansion as the world changes around us.
5. Make certain that records on individuals are given unique identifiers.

The reader's attention should be drawn to the fact that at no time was it recommended that race and marital status be abolished. Demographic data are imported into the occupational health and safety system from personnel records, and these data elements may be important to human relations, compensation payments, and government reports. The purpose of discussing them in the context of the occupational health and safety system is to encourage health and safety specialists to construct a meaningful database with data elements that contribute to health and safety decisions. Old habits are worth critical review when developing a new database. Each data element should be evaluated for its usefulness.

SELECTING DATA FOR CODING

Data selected for records should provide reliable information that can be grouped into a few salient subtopics. Each data item that is not an unique identifier, such as employee or agent ID numbers, must convey a precise descriptive feature about the topic of the record. We have already examined the ambiguities of using race and marital status to describe a person; and now we are ready to learn to code for better recordkeeping. The Incident Investigation record is an example.

We would like to discover where safety precautions need to be increased before additional incidents occur and the company loses more work time and more dollars. An incident requires a fact-finding investigation, which is followed by the safety committee's examination of the facts and a plan for future action—a perfect example of what recordkeeping is all about. The unique identifier is the incident ID number. A common data element in this record is the cause of the incident, which is typically broken down into unsafe acts and unsafe conditions. How does one describe an unsafe act or unsafe condition when there are so many possibilities? The answer is that you limit the choices and do not attempt to list every possibility.

The reason for limiting the choices is that the ultimate goal is to comprehend what the most common problems are and, from these, to develop a plan or program to curtail additional accidents. if there are a lot of causes, each with only one or two incidents, the ultimate numbers will be too small to reveal error patterns. Therefore, the size of the company and the normal number of incidents will influence the number of causes that can be handled effectively. A company with 1000 employees and only 5 incidents a year does not have be as selective as a company with 1000 employees and 200 incidents a year. The Incident Investigation in the second company is an important record containing critical facts.

A fact is one discrete unit of information. In its shortest form, a fact may be a single word, but for a computer it can be a select combination of a few words or a combination of words and numbers. To enable the computer to use facts efficiently, we need to reduce narrative text to a brief factual expression. As an example, we will begin with a narrative report and systematically reduce it to a coded computer entry.

> The accident occurred at *11:03 a.m.* in the *equipment storage room* when the *employee picked up an electric motor by its cord,* the *cord broke because it was frayed* and the *motor fell on the employee's left foot, breaking his big toe and severely bruising his instep.*

There are eight facts embedded in this story, all italicized. The employee is the first fact, and we will assume that he has an employee ID number that can be entered into the record. The second fact is the time of day, and we will assume there are data elements on the report for date and time. Additional data of importance that must be coded are:

1. equipment storage room
2. picked up an electric motor by its cord
3. cord broke because it was frayed
4. object fell on employee's left foot
5. broke his big toe
6. severely bruised his instep

Before coding, however, we will reduce the statements to a basic framework by asking a question: Who did something someplace at some time and why, and what was the outcome?

Who	Employee
Where	Equipment storage room
When	Date and 11:03 a.m.
What	Electric motor dropped
Why	Frayed cord broke
Result	Injured left foot

Every area of a plant must be identifiable by code. Some plants will have numbered grid coordinates for Title III reporting. Lacking this degree of detail, plants normally use building, department, and room numbers, which can be codes themselves. Equipment Storage Room should never have to be entered into the computer. A location code should return that description to the display screen following code entry.

The electric motor is a piece of equipment. It is the equipment that needs to be emphasized, not its specific type. The object that caused the injury just happened to be a motor, but if employees are trained to pick things up correctly, the training emphasizes *any* piece of equipment, not just electrical motors. If someone else picks up a heavy metal basket incorrectly and drops it on his right knee, there will not be one training course for picking up electric motors and another for picking up metal baskets. A second kind of data element can be added for this fact, which is action. There are two actions: the person's action of lifting the equipment and the other action which is free fall.

The last element is the injury, of which there are two parts: the part of the body injured and the nature of the injury.

A review of the data elements now shows:

Employee ID number	77889
Date of Incident	10/11/1988

Time of Incident	11:03 a.m.
Location of Incident	Room 105
Part of Body Injured	Foot
Nature of Injury	Broken toe, bruised foot
Object	Electrical equipment
Action	Lifting
Unsafe Act	Improper handling
Unsafe Condition	Damaged electric cord

If a standard practice in the occupational health and safety system is to use military time, there will never be a need for an additional field for a.m. or p.m. Military time is itself an effective way to code time. Room 105 should be coded in order to standardize data entry. Otherwise, one entry will read Room 105, another Rm 105, and still another R-105. Keep departments, buildings, and specific location data as separate fields, each with its own coding scheme.

Part of Body and Nature of Injury codes must match those used in the clinic and on the First Report of Injury. Standard codes like ICD-9-CM are so extensive that they go way beyond what the plant needs. Pertinent selections can be made from the list for regular use. A tape of ICD-9-CM codes is purchasable for computer use, but this is excessive since the majority of the codes will never be used. It is also reasonable to construct the plant's own codes.

Objects, actions, unsafe acts, and unsafe conditions are all codes to be developed for the plant. To code objects, begin with all the general categories of objects that can cause injury, such as tools, equipment. furniture, doors, windows, and vehicles. Break these down to read Tools, hand, Tools, power, Equipment, electrical, Equipment, mechanical, and so on. The same holds for Actions, Unsafe Acts, and Unsafe Conditions.

Codes are easiest to enter reliably and to read readily when they have meaningful abbreviations. The same Incident Investigation tables may be usable for all four of these categories by applying letters to numbers. All actions may begin with AC, all objects with OB, unsafe actions with UA, and unsafe conditions with UC. When data are grouped, tools of several types and equipment of several types always provide room for growth. For instance, this list of codes might read:

OB 1001	Tools, hand
OB 1002	Tools, power
OB 1101	Equipment, electrical
OB 1102	Equipment, mechanical

A real Incident Investigation record will contain a lot more data than this, but the examples given should provide some insight into how to shift from narrative text form to a coded form more suitable to computers. Proper screen form design will always include a field for additional narrative explanations of the principal facts.

When data are retrieved for reports, they should be from the coded fields. Narrative comments should be reviewed for their pertinence but generally avoided in reports. A safety committee review should evaluate the basic facts before relying heavily on comments. If there appears to be a pattern of lifting accidents, then more training on lifting may be needed. If there are numerous unsafe conditions with damaged electrical cords, more attention should be paid to reporting and following up on electrical repairs. The essential point to bear in mind is that the coded responses will keep attention focused on larger relevant issues rather than specific detail, but when the specifics are needed, they can be made available as a follow-up, possibly to detect underlying causes.

In many plants, there are entries for the kinds of data elements given here. My experience is that only about half or fewer of these are coded. Where the recorder is given freedom of expression, it is invariably more difficult and more time-consuming to evaluate and report incident data. It semed reasonable to select these fields in an Incident Investigation to show how to convert from text to coded data because the data are familiar to almost everyone in occupational health and safety, but the same method can be applied to any occupational health and safety record.

Using a group of old narrative records, the major steps are:

1. Underscore only the facts in the text.
2. Construct a relevant framework including factors like Who, When, Where, Why, What, and the Result.
3. Fit the facts to the descriptions.
4. Reduce the facts to general descriptions.
5. Add codes to the general descriptions.

READY-MADE CODES

The medical community has been active in coding data for many years because of their great need for fast retrieval, statistical analysis, and epidemiological study. They have recognized the importance of describing medical diagnoses in a standardized manner. Nearly a dozen different systems have been developed, and while none are directed at occupational health and safety, some portions of one or two of them may be adapted by health and safety persons.

The development of these coding systems has been driven by the acute need of hospitals and clinics for standardized descriptions of anatomic sites, diagnoses, sources of injury, types of injury, severity of injury, and other medical data. Each category is suited to a particular environment. A brief discussion of prominent coding systems follows:

American College of Radiology (ACR) was developed in 1955 to simplify filing and retrieving. This code system preceded the use of computers.

Standard Nomenclature of Diseases and Operations (SNDO) was based on anatomic site and disease etiology. The fifth Revision occurred in 1961 so this is also a precomputer system.

Systematized Nomenclature of Pathology (SNOP) was computerized and depended on punch card entry. It is based on anatomic site, morphology, etiology, and chemical/clinical changes.

Systematized Nomenclature of Medicine (SNOMED) evolved from SNOP in about 1971 and uses numerical codes. It works well for nursing reports and health care facilities. The code consists of a letter followed by five digits and was not designed for computer use.

International Code of Diseases (ICD), which has undergone 9 or 10 revisions, is generally referred to as ICD-9-CM. The ninth revision was completed in 1978 and is the most used. CM stands for Clinical Modification. Its purpose is to categorize medical terms rather than to define them accurately and it is based on anatomic site, etiology, morphology, and function. Each of these topics has its own key. This coding system is used for diagnoses and has a five digit code. It is extensive with 16 broad classes of disease and injury that break down into more than 10,000 categories. The complaint is that the code descriptions permit too much opportunity for error.

Abbreviated Injury Scale (AIS) is used for anatomic injury and works together with ICD.

CPT-4 is used in physicians' offices but is not adaptable for hospitals and health care facilities. It may have some usefulness for nurse-operated plant clinics.

Severity of Illness Index (SII).

Computerized Severity Index (CSI).

Diagnosis Related Groups (DRG).

Each of these systems was designed to answer the needs of a particular kind of recordkeeping. Thus, some are more office oriented, where financial records depend upon specific diagnoses. In the occupational health and safety setting, ICD-9 and CPT-4 codes may be excerpted for use, provided that the codes are judiciously chosen.

Much of the data that are entered for industrial hygiene are measurements made during sampling, which is not codable material. Industrial hygiene does have some ready-made codes, but not in systems like those of the medical community. Laboratory analytical methods largely depend upon established methods which are already numbered and can be used in a computer system. NIOSH methods all have numbers. The Intersociety Committee (ICS), which consists of the American Chemical Society and eight other prestigious organizations, also has numbered methods for laboratory analysis of samples. These numbers are actually codes and can be used without alteration, provided that they are accompanied by their source. To indicate the source of the code number, simply precede the number by a letter or two representing the organization.

GLOSSARY

Acceptance. Official adoption of the completed system by the users after thorough testing and assurance that the developer has completed all contractual agreements.

Access control. Access to the computer can be controlled by action-oriented means such as limiting user read or write privileges, or by object-oriented means such as limiting user access to files and fields.

Ad hoc data retrieval. Specific data search done on a one-time basis.

Algorithm. A set of rules written to accomplish a specific task in a minimum number of sequential steps.

Analog signal. The type of electrical signal that is produced by a voltage measurement. The number of possible values is infinite, since voltages are not required to be an integer number,as is a digital signal. (See *digital signal.*)

Analysis. This study of facts to determine the salient points and relationships that may illustrate a hypothesis or concept follows selection of the data during review and precedes data interpretation. This step may include statistical testing.

ANSI (American National Standards Institute). This group establishes national standards for computer languages and computer equipment.

Application generators. Cost-saving, fourth-generation language tools that use generic routines to automate code generation, such as screen painters and report writers.

Applications software. Programs written to carry out a special purpose.

Archives. Data which are separated from the working database for long-term storage in a location different from the capturing computer, but which can be retrieved as needed.

Array. A series of items arranged in a meaningful matrix of n dimensions, each item having identical relationships to the categories represented by the dimensions.

ASCII (American Standard Code for Information Interchange). An information code used in most computers to represent letters, numbers, and symbols which facilitates information exchange between equipment of different manufacturers.

Assembly language. A second generation programming language that uses abbreviations and symbols to represent a function. Must be converted to machine language to be executable.

Assessment. An appraisal or evaluation of something that gives it a fixed value or quality.

Attribute. The value associated with an entity that describes the property of an object in the database. In relational databases, attributes are fields that become the columns of the relational table.

Audit trail. A computer program used in data processing systems to record a step-by-step record of all the processing of a transaction, beginning with the entry of the original source data. This operation is transparent to the user.

Auxiliary storage. Magnetic media devices that are used for long-term storage of data and stored separately from the computer.

Backing out. The process of retracing all menu selections in reverse sequence in order to reach the lead menu for a new selection. This becomes necessary if the menus are organized hierarchically.

Backup. An extra copy of a program or database that is made frequently to guard against a disastrous loss in which all the data may be lost.

BASIC (Beginner's All-purpose Instruction Code). A common high-level programming language that was originally developed at Dartmouth University for academic use but that gained popularity with microcomputer users.

Batch processing. Processing of data with identical requirements as a group, which can be done at off-peak times for improved system efficiency.

Binary. A number notation system in which each symbol represents the power of two, having only two digits, 0 and 1. All values can be expressed in a combination of these digits. This is in contrast to decimal digits that can be expressed in any of 10 values, 0, 1, 2, 3, 4, 5, 6, 7, 8, 9.

Bit. A binary unit of data that can exist in one of two possible states, 0 or 1. A bit is the smallest unit of data.

BMDP. BMD Biomedical Computer Programs, P Series, for applying statistical techniques to medical data analysis.

Boolean operator. A character that indicates an operation such as Equal To (=), Greater Than >, or Less Than <

Boot up. The process of loading a microcomputer with its operating system program, which brings the machine into operation ready to load the user's software programs.The reference is to "pulling oneself up by his own bootstrap."

Box head. The row label that describes the contents of rows in a tabular report.

Browse. A function of some programs that permits the user to view the contents of the database, screen by screen, for review and editing.

Bugs. Mistakes (errors) in logic or syntax in a computer program that cause the program to fail in function or reliability.

Byte. A group of 6, 8, or 9 bits treated as a unit. This is the smallest addressable unit in a computer. Most bytes consist of 8 bits.

Calculations. Planned algorithms or methods for the solution of mathematical problems.

Canned query. A data retrieval request that can be saved and stored for repeated use.

CASE (Computer-Aided Software Engineering). Fourth generation language tools for system design that automate chart and diagram production with great savings in time and costs. Some CASE tools provide analysis features in addition.

Cassette tape. A method for storing computer data that is similar to audio cassette tapes, but that has slow speed and low capacity compared to other electronic storage media.

CCITT (Comité Consultatif Internationale de Télégraphie et Téléphonie). A committee that sets standards for international communications.

Centralization. Concentrating computer power in a single location.

Central processing unit (CPU). The functional part of the computer system that reads and interprets instructions for carrying out the processing of data.

Character. Computer symbols having two equivalent forms, one understood by humans, such as the letters, numbers, and symbols on keyboards and screens, and the other understood by the computer in the form of bits and bytes.

Chart. A graphic representation of data, usually in a tabular form but sometimes in a form that shows curve plots.

Chip. See *microprocessor.*

Circuit card. Electronic circuits are etched on thin, rectangular fiberglass or epoxy cards that are coated with copper. The circuitry is etched in copper, and the card is inserted into the computer. This facilitates upgrading of the hardware system with the least disruption to the system.

COBOL (COmmon Business-Oriented Language). A high-level programming language that was developed especially for business applications.

Code. Instructions for computer operations written in a programming language.

Coded data. Data that have been assigned an abbreviated designation in order to standardize the database and to save storage space.

Coding 1) The process of writing programmed instructions for a computer. 2) The process of assigning groups of letters or symbols to represent specific objects or descriptions.

Command. A computer instruction in the form of a programming language that instructs the machine to carry out a procedure consisting of one or more operations.

Command language. See *high-level language.*

Communication. 1) Imparting or transmitting knowledge. 2) The means by which a computer sends data to, and receives data from, other computers.

Communication system. A system that transmits information encoded by a sender to a receiver, who decodes the information.

Compile. Translation of programs written in high-level languages understandable to the programmer into machine- level language understandable to the computer by a specially written software program called a *compiler.*

Complex report. A multipartite report that may be lengthy and that may carry numerous breaks for summaries throughout the report.

Computation. The process of carrying out planned calculations.

Computed variable. A field that has a calculation attached to it which transforms the value of the field when the data are processed.

Confidentiality. A special concern for sensitive information in order to minimize possible invasion of personal privacy.

Configuration. The arrangement of the parts of a system and their relationship to one another, including the peripheral devices that are attached to the computer system. Both hardware and software are configurable.

Construct. A statement in a computer language that directs the machine to follow a specified procedure. High-level languages have more complex constructs than low-level languages.

Control break. The point in a report at which the specified field shifts its identity. For example, if Department is the field on which the control breaks occur, a break would appear in the report between Department A and Department B.

Control key. A keyboard key labeled CTRL that is used together with other keys to cause something to happen for which there are no specified characters. Usually shown in written directions by an uppercase carat. (e.g.,^A for CTRL-A)

Cost/benefit analysis. The analysis of costs for a proposed project that shows the benefits to be derived that will justify the cost. Both costs and benefits must be shown in identical monetary units.

Cost-effectiveness analysis. A project cost analysis based on comparable costs for alternative ways to achieve the same approved goal when it is impossible to attach an explicit cost to each part of the project.

CRT (cathode ray tube). A television-like screen for the display of data. (See *video display terminal (VDT)*.)

Cursor. The screen indicator that shows where the computer expects the next character to be entered. An important part of the user interface.

Daisy wheel printer. A high quality impact printing method with a rotating wheel of raised characters that produces letter quality output.

Data. Facts representing measurements, observations, or financial figures.

Data analysis. Detailed exploration of all the data used in an application to identify each of the elements included and their relationship to one another. This is a controlled method for handling data.

Database. A collection of related data on the same subject formally organized in a logical order.

Database administrator. A person responsible for maintaining database integrity who defines, organizes, and protects the database and controls database access.

Database design. The definition and organization of data elements and data relationships into records for the support of an application.

Database management. Control of a database system to ensure user adherence to standards protecting the consistency and reliability of company data.

Database management system (DBMS). Set of programs used to establish a database and to control access to and organization of data in the database. These systems typically have accompanying query languages for ad hoc data retrieval.

Database manager. See *Database administrator*.

Data consistency. The standardization of data for storage.

Data definition language (DDL). Language that is part of the database management system that defines files, variables, and their relationships.

Data dictionary. A part of the software that documents each of the data elements in the database, defining their meaning, format, length, and relationship to one another.

Data dump. An ad hoc request that all the data be output for a requested field(s) in the database.

Data element. The smallest unit of a database record. See *field (variable)*.

Data entry operator. An individual whose principal assignment is to enter data into a computer from source documents.

Data input. Entry of data from original source documents into the database.

Data integrity. The assurance that data entering the system will not differ from their source form. Also database integrity.

Data interpretation. See *interpretation*.

Data loggers. Small machines with a computer chip that can process data and output the results on a small paper tape.

Data manipulation. Processing of data preparatory to storage by sorting, performing computations, and making comparisons.

Data manipulation language (DML). The data language used to transfer data between the application program and the database.

Data retrieval. The search for specified data located in database storage bringing them into computer memory, after which they are displayed on the screen or sent to an output device.

Data transformation. The alteration of data in order to store them or see them in a different form. An example is to convert a number to its logarithmic equivalent.

Data validation. The assurance that database integrity will be maintained through methods such as completeness checks and validating totals during batch processing.

Data verification. The assurance that data entering the system are compatible with data specifications by using range checks, checking for blank spaces, checking the field format, and other methods.

Debug. The detection and elimination of sources of error in programming routines.

Decentralization. Computer power spread over two or more locations.

Decision making. Reaching a rational conclusion or making a choice based on information received.

Decision support system. A fifth generation language product that derives decisions based on the knowledge flow as opposed to a data management system that provides information for users to make their own decisions.

Default variable. A field having a prespecified value that is used in the database unless the user overrides it with a different value.

Delimiters. Characters that separate and organize elements of data into groups (e.g., slashes between dates).

Descriptive statistics. Simple statistical tests that attempt to show measures of central tendency like mean, median, mode, and standard deviation.

Development. Software construction that covers several phases such as systems analysis, system design, programming, and documentation.

Digital signal. An electrical signal that pulsates between two conditions, *ON* (one) or *OFF* (zero). Combinations of these signals represents alphabetic and numeric characters.

Disk drive. A high-speed rotating spindle housed in a box. A diskette is inserted and spun while recording and retrieving data.

Diskettes. See *floppy disks*.

Disk packs. Storage medium for computers in which metal platters are coated with a metal oxide for magnetic media. A series of platters connected to a hub.

Distributed database. Portions of a database housed at remote locations where the data are most frequently used.

Distributed system. The distribution of computing power and data using minicomputers or microcomputers at locations where the processing is required. A network of computers linked through public utility communications lines in which processing is done on individual machines.

Documentation. A written record of the project that establishes design and performance descriptions for use by users and systems personnel. This is a detailed account of facts about the programming and system operation given in supporting materials and within the program itself.

Domain. The set of related data that represents all the possible values of an attribute.

DOS (disk operating system). An operating system developed by Microsoft for IBM and compatible microcomputers.

Download. The transfer of data from a larger computer environment to a smaller computer environment.

Dot matrix printer. A typical microcomputer printer in which the characters are formed on paper by closely spaced dots. There are different grades of print quality that can be expected of these printers.

Electronic files. Data stored on space-saving magnetic media.

Emulator. Software that makes a terminal of one manufacturer look like the terminal of another manufacturer.

Encryption. Security protection by cryptographic coding of files; where intensive security is advised.

Enter Key. See *Return Key*.

Erroneous data. Test data that have been deliberately distorted to mimic errors in user input and that the system should reject at data validation.

Error check. Any method used to check incoming data for inaccuracy.

Error message. A message displayed on the status line to indicate to the user that a system problem exists. Users should respond to the message before proceeding further.

Exception reports. Periodic or occasional lists of deviant data that are used to located unusual conditions.

Export/import. Files that are sent from, or taken in by, a system through a programmed interface.

Exposure group. See *homogenous exposure group (HEG)*.

Feasibility study. Phase of project initiation that describes the user's problems and suggests a tentative approach to their solution.

Feedback. In an interactive system, the video display screen shows the machine instructions that the user gave through the keyboard to verify how the machine is expected to respond.

Field (variable). A unit of data within a record that contains only one item of information and has a specified location within the record.

Field length. The number of spaces needed for the field to contain the data.

Field type. The form of data that is needed for input, such as alphanumeric characters or integer or real numbers.

Fifth generation languages. Incoming generation of computer languages that emphasize artificial intelligence and expert systems.

File. An ordered collection of fields which is stored as an identifiable unit on a disk or a tape. Also called a *data set*.

Filter. A user's request that is programmed with a retrieval to suppress all data except those which meet the range or other data specifics.

First generation language. Machine-level language that is given in binary form.

Floating point arithmetic. Computer arithmetic in which the machine keeps track of the decimal point's position.

Floppy disk (diskette). A thin, flexible mylar disk encased in a semirigid protective jacket. This is the storage medium common to microcomputers; it is economical and easy to use but has relatively low speed. Come in diameters of 3.5, 5.25, and 8 inches.

Flowchart. A programmer's tool for displaying a sequence of operations through an established set of symbols and other marks that represent steps in the functions of a computer program.

Format. Method of prescribing the arrangement of data.

Forms. Organized outlines for capturing specified information.

FORTRAN (FORmula TRANslation). A high-level language designed primarily for scientific applications.

Fourth generation languages (4GL). High-level languages that provide automated programming tools for generating screens, reports, and system design.

Function keys. Special keys on the keyboard that perform useful operations for the user.

GIGO. A term meaning "Garbage in, garbage out" commonly used to describe the unreliable information gained from a computer when close attention is not given to data placed in the system.

Government report. Usually mandated to be produced annually for state, federal, or local government agencies.

Graphics. A pictorial representation of data or information that can be displayed on the screen or output on a peripheral device like a printer or a plotter. Often drawn and controlled by a hand-held device called a *mouse*.

Hard copy. Printed output, as opposed to soft copy which is displayed on the video display screen.

Hard disk drive. A microcomputer box containing a spindle that rotates coated magnetic disks for electronic storage.

Hardware. The equipment that makes up the physical part of a computer system, as opposed to the written programs that direct the system operations.

Hierarchical database. Fixed relationship between the data elements in which there is a parent field to which all additional fields are related, as a child is to a parent. This necessitates navigating the database through parent field relationships.

High level languages. Third-generation languages that are written in English-like statements where each statement corresponds to several machine instructions. These languages must be compiled into lower level machine language for computer instruction.

High speed printer. Used with mainframes and large minicomputers because it can output data at speeds in excess of several hundred lines per minute.

Homogeneous exposure group (HEG). A group of employees exposed to the same agents at the same approximate rate of exposure.

Host computer. A computer attached to a network that provides the primary functions of computation, database access, and special programs for several or more users. Archived data are sent to the host computer for long-term storage.

Implementation. Installation of the new system of data management and conversion from the old system to the new one.

Index. A duplicate file that has been sorted according to a specified key for rapid data access at retrieval.

Information. Data that have been logically grouped and organized into meaningful knowledge.

Information engineering. A database management type of software engineering that shows the project in its corporate setting and charts data elements and the flow of data in the organization.

Information services. A group of people or a department within a company that controls and oversees all computer services.

In-house. Within the company.

Input. Data transferred from an external source to the computer that are checked and edited upon entry to preserve accuracy and validity. These are the data that the program read and process.

Input specification. User definition of how data should look when they are entered into the database.

Installation. Hardware and software placed in position and made operable.

Integrated database. A database that is used for several kinds of data that can be shared by different users. Users can be restricted to the specific data they need to share.

Integrated electronic circuit. See *microchip*.

Integrated software. Two or more types of software that work as one to provide the user with more than one service, such as word processing and spreadsheet analysis in one package.

Integrity. The extent to which the data processed are reliable, accurate, and free from error. In a system, this refers to the proper function of hardware and programs, and the physical security and safety of the database.

Interactive. A system that permits communication between computer and human.

Interface. The boundary between two entities, such as the human and the computer, or between a software program and the computer machine.

International Labor Organization. An organization of 150 participating member nations that strives to improve labor and living conditions for working people by establishing labor standards and disseminating information.

Interpretation. Following data analysis, a subjective evaluation of the validity of the facts as presented to determine that they support a hypothesis.

Interpreter. A high-level language that resides in computer memory with the application program and translates the program to machine-level language, statement by statement, as the application instructions are given.

Iterative. A repetitive process, especially the repetitive review of screen prototypes that provides users with the opportunity to revise and modify screens for data capture at low cost to the developer.

Join. The combination of two relational tables through a common field so that all data retrieved match across the search field.

Jump. An instruction to the machine to bypass some instructions and cause the program to proceed to a different location before continuing.

Keyboard. Input device that is like a typewriter keyboard. As keys are pressed, ASCII signals are sent to the computer system, providing instructions to the computer for accepting or processing data.

Knowledge. The comprehension and understanding of an organized body of information.

Laser printer. A letter quality printer used with microcomputers that has a semiconductor laser as a light source for its electrophotographic recording system.

Linkage. A mechanism for connecting one program area to another to facilitate navigation through the system.

Live data. Data taken directly from a user's source documents.

Local area network (LAN). A computer network spread over a limited geographical area that is connected by coaxial cable owned by the users rather than by communications lines owned by a common carrier.

Logon. The procedure for a user to be identified for access to a program in which the system checks the user's identity and password, records the date and time of access, and initiates the program. This procedure is separate from operation of the application functions.

Logout. The procedure for signing off of a computer program when the user has completed work, at which time the system records the date and time of departure.

Lookup. See *table lookup*.

Mainframe. Very large computers capable of processing billions of characters of data at very high speed.

Maintenance. Making adjustments to meet the changing demands on existing software which is properly functioning but in need of upgrading and enhancement.

Management reports. May be periodic. Contains cost, personnel information, and production data.

Manual entry. Entry of data on paper forms.

Matrix. See *array.*

Memory (RAM). Temporary storage area in the computer for holding a program and data while in use, which is lost when the computer is turned off. Temporary storage increases the speed with which data can be written to or read from the computer.

Menu. A computer screen display that provides a set of choices for action and that divides the system by functions linked to work practices.

Menu driven. All of the available options for selecting a path through the system can be made from menus, and the user needs no other source for system direction.

Methodologies. Sets of sequential rules and procedures that can be used to solve problems in engineering disciplines.

Microchip. A silicon wafer containing transistors, diodes, and other circuit components interconnected by etched copper channels.

Microcomputer. A computer configured for a single user that includes keyboard, video screen, memory, and computational elements.

Microprocessor. One of the principal components of a microcomputer. See *microchip.*

Minicomputer. A computer configured for simultaneous use by a small number of people that is larger and has more computing power than a microcomputer.

Models. Diagrams and plans that emulate a complete system design or database design and that are used to predict success or failure by early detection of errors.

Modularization. Minicomputers and programs constructed in units of limited size to simplify development and maintenance.

Module. A unit of limited size in a minicomputer and program.

Monitor. A video screen device that displays (monitors) computer input and output. See also *video display terminal (VDT).*

Monochrome. Video screen display of single-color characters against a dark background; the combination can be reversed. The colors are limited to amber or green.

MS-DOS. A disk operating system for IBM and compatible microcomputers produced by the Microsoft Corporation.

Multilevel help. Help screens that provide assistance through several hierarchical levels of definition. The main topic will offer several choices of subtopics, and each subtopic may provide one or two lesser topics for description.

Multipart profile. A profile report consisting of several or more sections, each contributing one type of information on the report subject.

Multitasking. Operating software in newer microcomputers that controls two or more programs in memory at any given time.

Multiuser system. A computer system with several or more terminals or computers that are simultaneously operable. A network is an example.

Navigate. To follow access paths through a database as defined by the database design structure.

Needs assessment. A study of the user's computer system requirements that can be combined with a feasibility study.

Network. A group of computers linked together by communications lines. See *local area network (LAN)* and *distributed system*.

Network database. a database designed so that each of the elements is linked to the others through pointers. Some of the elements are termed *owners*; others are termed *members*. Each member may have more than one owner.

Nonprocedural languages. Programming languages.

Notification. A report in which a form letter notifies someone about a result or concern.

On-line. A condition of direct interactive communication with the central processing unit of a computer.

Operating software. The software that controls and directs the computer and execution of the programs in an orderly operation of the system.

Operating system. Group of programs that controls the computer and integrates its mechanical functions.

Operations manual. Software or hardware documentation written in the technical language of computer systems for systems personnel that describes operating commands and procedures, software installation instructions, file recovery procedures, and backup procedures.

OS/2 (OS/2 Extended). New IBM operating system for PS/2 systems.

Output. Transfer of data from storage and required processing to an external form such as a video display screen or a printed page.

Overhead. Programming and system demands that increase the amount of processing required beyond normal expectations.

Paper files. Records maintained on paper and stored in file cabinets.

Parallel processing. The simultaneous operation of the new computer system and the old method of data management while making an orderly transfer to the new system. As the new system proves reliable, the old one is gradually dropped.

Parity check. The addition of noninformation bits to data to bring the total of all bits to an odd or even number, depending upon the specification of odd or even parity. This is an error detection technique for data communications.

Partitioning. Dividing the database so that access to data is controlled by specific users. Partitioning also applies to division of storage in the computer wherein more than one operating system can be used.

Pascal. A high-level computer language that is user oriented and very structured in its syntax and use.

Peripheral equipment. Hardware that is linked to the computer, such as disk drives and printers. These units work in conjunction with a computer but are not an integral part of it. Other examples are the plotter, tape reader, and analog to digital converter.

Personnel identification. Methods of identifying authorized computer personnel through fingerprints, hand shapes, voice patterns, eye retina patterns, and other similar human characteristics.

Physical security. Prevention of physical destruction to computer equipment (hardware and software) or a database by man-made or natural disasters.

Porting. Moving software from one manufacturer's computer equipment to another or from one kind of a computer to another (e.g., minicomputer to microcomputer).

Printer. Peripheral equipment that outputs data from the computer to paper, producing images of text and sometimes graphics.

Privacy. A legal right of individuals and organizations to protect their data from being viewed by unauthorized persons.

Procedural language. Specifies how data are retrieved and output by telling the system what to do with the data.

Procedure manual. Documentation that describes the actions to be followed by users or systems personnel in order to ensure maximum benefit from the system.

Processor. See *central processing unit (CPU).*

Production reports. Reports that track administrative information and are generally produced at regular intervals.

Profile report. A special report on an employee, agent, or work location that shows a list of related data on the subject at any single time or in a time series.

Program. A series of commands written in a language that can be translated into machine language to direct computer procedures.

Programming. The act of writing programs in a language designed for computers.

Prompt. A screen display indicator showing that the computer will accept a specific kind of data or other instruction that the user is to supply through the keyboard.

Proprietary. The exclusive right of a manufacturer to sell, copy, or use a program that can be delegated to another user according to the manfacturer's restrictions.

Prototyping. A method of developing applications software in which the developer provides iterative screen designs for the user to review and change before finalizing the product.

Pull-down. Screen technology that scrolls downward an inset window with information in it. Typically used for menus, especially when there are series of menus (nested menus).

Qualitative assessment. An evaluation which is made with insufficient valid data for quantitative measurement, but which can be subjectively characterized by some property or descriptive definition.

Quality assurance (QA). Standards and procedures devised for the express purpose of assuring the product will always meet established criteria.

Quantitative assessment. An evaluation that can be objectively measured in numerical terms.

Query language. A high-level computer language used for ad hoc data retrieval and data modification which is highly adaptable to users. Query language is usually provided by a database management system for easy access to the database.

Query paths. Navigational pathways through the database set up by a query language system.

Questionnaire. A form with many questions to be answered.

Random access memory (RAM). See *memory (RAM).*

Read. Process of taking in data from an external source and transferring them to internal storage. See *write.*

Real data. Good, valid data that do not necessarily come from the user's source documents.

Real number. Any number that is expressed as a decimal number and that may be either positive, negative, or zero in value.

Real-time. Performance of a computation during the actual time that the computer is monitoring and recording a physical action in the process equipment to which it is linked so that the process can be regulated and controlled.

Record. A collection of fields related to a specific kind of information in the database which is also a unit of storage.

Recordkeeping. The preservation of data in an organized manner for eventual retrieval and use as recorded information.

Record locking. A problem for some multiuser systems when more than one user attempts to access and store the same field.

Recovery procedures. See *rollback recovery.*

Redundancy. The duplication of data elements in the database.

Relational database. A database design that organizes data in columns and rows, referred to as a *table*. The *database* is a collection of tables that are related to one another through common key elements for efficient search and retrieval.

Reliability. The dependability of a computer system such that the information used in decision making is known to be accurate and consistent.

Repair. Detection and elimination of bugs in software.

Report. A collection of data output by the computer system that provides information to the user.

Report writer. A fourth generation software program that automatically generates source code to format database output by the specified sort order.

Requirements analysis. See *needs assessment.*

Response time. Time required by the system to react to an input stimulus which is measured as the period between the input of the last character and the receipt of the first character of the reply.

Return Key. When pressed, this key terminates the displayed line and returns the cursor to the beginning of the next line. The equivalent of a carriage return on a typewriter.

Review. Selection of data for a critical appraisal of their value to a hypothesis or context. The selection is then subjected to an analysis, usually by statistical means.

Right justify. A printing direction that keeps right margins even by spreading the words across the allotted width of the page.

Rollback recovery. The process of reversing recent actions of a downed computer system to restore some or all of the database to its reliable condition at a previous time. This protects against loss of the entire database following an emergency or a disastrous computer failure.

RS-232C. A standard type of cable connector used in computer systems.

SAS (Statistical Analysis System). One of the most popular of statistical analysis packages that is frequently used in medical data analysis.

Schema. A database outline that specifies the properties of each data element and the relationships between data elements and records in the database.

Screen design. User-friendly screen display that blends together human, hardware, and software features for interaction.

Screen display. The input or output techniques that reveal the operation of the computer or the picture of the database on the video display screen.

Screen painter. A fourth-generation language tool that permits the screen to be designed and organized according to the requirements of the user.

Scrolling. The capability to move lines on a screen display up or down and right or left by the action of one key. Scrolling can take place line by line or by a block of lines at a time.

Second-generation language. Symbolic language that uses letter symbols to represent instructions to the computer and translates them into machine language on a one-to-one correspondence with computer instructions. Specific to a microprocessor and commonly called *assembly language*.

Security. Protection of a computer system and data against any accidental or intentional damage and against unauthorized disclosure, modification, or destruction.

Security administration. Authority delegated to a person to grant or revoke computer privileges, control passwords, and establish and review security procedures.

Shared data. Data that can be accessed by more users than those in a single discipline. This does not imply that all data can be shared; there is some access control in order to protect confidentiality.

Simple report. A report usually on one topic, short and lacking control breaks, which may be issued on a one- time basis or periodically.

Software. A set of computer programs, written procedures, and associated documentation that provides instructions to the computer as it resides in memory during use.

Software copyright. The legal right of the developer to be the sole owner of a software product; identical to a publisher's copyright of a book.

Software engineering. The analysis, design, and development of software and its programming.

Source code. The program of a language other than machine language which is not executable until converted into machine language. Some software permits the user to obtain source code, but most software manufacturers protect the software from user access.

Special report. Ad hoc or formal presentation. Special information provided on a one-time or infrequent basis.

Spreadsheet. A special application software package used for viewing numerical data, and which may process the data by manipulation or number crunching. Especially useful for viewing data under"what if" conditions.

SPSS (Statistical Package for the Social Sciences). A well-known statistical analysis package.

SQL (Structured Query Language). A language developed by the IBM Research Laboratory in San Jose, California, under the name "System R" that is based on relational database concepts and used to retrieve data from the database.

Storage. Transfer of data and programs from computer memory to a place on storage media, where they can be accessed when needed. Storage media may be floppy disks, disk packs, magnetic tape, or hard disks.

Standard report. Information provided on a regular basis.

Status lines. A feedback mechanism for interaction between the computer and the user that contains a variety of information such as error messages, names of programs or modules, special characteristics of fields, and function key availability.

Structural programming. Programming done by decomposition of the program design into segments called *modules*, each of which has a given function. This provides a sound basis for future repair and maintenance. See *modularization*.

Stub. Headings that describe the contents of columns.

Subset. Data selected from the database and placed in new files prior to export or used for temporary purposes.

Syntax. The rules governing the structure of a language.

System. A series of related processes or equipment linked to perform a specific task with an orderly arrangement of components or parts into a connected and interrelated entity.

System administrator. A person who is responsible for protecting and overseeing the operation of a system and who may have a staff with assigned tasks.

System architecture. The basic system design that has an organizational context around which all other details must be designed to fit.

System design. A detailed description of the technical and other specifications that make the system operational.

System operation. Computer functions necessary for running applications software, which are carried out by systems specialists rather than users of minicomputers and mainframes.

System security. The prevention of incorrect or lost data because of system failure or user errors, including database security.

System specifications. A permanent record of the structure, flow, and procedures of a system which can be used to test and evaluate the system at implementation time.

Systems analysis. The process of analyzing existing systems to evaluate possible improvements in methods and procedures for data management.

Systems analyst. A person who designs and develops applications systems for a computer.

Systems manual. Documentation for systems personnel who manage the application software, which is used by the database administrator and written in language familiar to database experts.

Table (file). Relational database structure in which there is a two-dimensional array of data with columns as fields and records as rows.

Tables. Tabular-designed reports with columns identifying topics and rows supplying data to fit those columns.

Table body. The part of the table that includes all the data, but not column or row headings, the title, or footnotes.

Table lookup. Data retrieval from a dictionary file for a specific value to be used in a form. The process is built into the system, and helps to standardize data and save users data entry effort.

Tape cartridges. Devices similar to audiotape cassettes that record electronic data and are used for long-term storage or transport of data to other computers.

Terminal. A video display screen and keyboard that is linked to a minicomputer or mainframe.

Testing. Trials for determining the reliability of software and hardware operation through an established set of computer directions. It verifies that data are being correctly accepted, processed, retrieved, and output.

Text. Passages of narrative writing.

Third-generation languages. High-level languages such as COBOL, FORTRAN, BASIC, Pascal, and C that are written in English-like statements and must be compiled in order to be translated into machine language for execution.

Tickler file. A file maintained for reminder purposes.

Time-sharing. Computer system for which time is purchased by several or more users who require a specific application database that they can use over standard communications lines and modems.

Top-down approach. A programming technique that proceeds from the general to the specific and that builds quality assurance into the program by systematically checking the structure as program development continues.

Trainer. A person who is knowledgeable about a computer software program and capable of training users in the skills necessary for use of the program.

Training. A special type of education in which the learner is taught skills to be used that can become repetitive, as opposed to teaching, which emphasizes reasoning processes.

Training manual. Documentation specially written for users to help them gain the skills needed to use a computer software program.

Trash data. Unrealistic grab bag data that have no meaning in the context of the database and should not be used to test a new system.

Tutorial. Step-by-step written directions for users to follow in learning how to use a computer software program.

UART (universal asynchronous receiver/transmitter). A computer connector that converts parallel data bits into serial form or serial data bits into parallel form for transmission.

UNIX. An operating system for computers developed by AT&T.

Update. The processing of additions, changes, and deletions to the database that provides the latest data available.

Upload. Transferring data from a smaller computer system to a larger computer system. See *download.*

User friendly. A system that interacts well with users and does not intimidate them. The user is guided through all operations in a language that both the human and the computer can use.

User identification products. Devices that are employed for user access to a computer system, such as authorization identifiers like passwords. Commercial products also are used for identification procedures.

Users. Persons who utilize the output from a computer system in their daily activities and who store their data in electronic files.

User's manual. Documentation designed and written for easy user comprehension that can be used as a permanent reference source.

Variable. See *field*.

Vendor. A purveyor of commercial hardware or software products.

Video display terminal (VDT). A television-like screen that acts as a computer monitor and displays characters and other machine-related information to the user in combination with a keyboard.

Virtual memory system. A method of using memory capacity frugally that gives the user the impression that there is no memory limitation. Data and parts of programs are held in auxiliary areas, and are brought into active memory when needed.

Wild card. A retrieval request that is basically nonspecific and only partially defined. Retrieval produces all records, beginning with the specified criteria.

Winchester disk. See *hard disk drive*.

Windows. Divisions within the computer screen that contain information from several places in the system, permitting the user to view all the information simultaneously.

Word processing. The writing and manipulation of narrative text under the control of applications software designed for text composition, formatting, editing, and printing.

Work practices. Normal user work routines in the workplace that need to be considered when building a computer system in order to provide compatibility between computer use and typical office operations.

Workstations. Independent areas of computer control in a network of computer operations that provides the user with both a personal computer system and the advantages of access to other network operations.

Write. The process of sending data to a database for storage in a specific location on magnetic media or to an external device for output. See *read*.

Xenix. An operating system.

BIBLIOGRAPHY

Abdelhak, Mervat. 1980. Toward a definition of information: functional and logistical perspectives. *Medical Record News.* 51(3):105–107.
An interesting discussion about the real nature of information systems and how our interpretation of the word information affects the manner in which we plan, develop, and evaluate information systems.

American Medical Association. 1984. Guide to Evaluation of Permanent Impairment.

American Petroleum Institute. 1983. Hazard communication rule. *Federal Register.* 48(228)Nov.25,1983:53300.

Aron, J. D. 1969. Information systems in perspective. *Computing Surveys.* 1(4):213–236.
Although this paper was written before database management was a widely adopted concept, it is still relevant in the larger context of information systems that are needed to conduct business, and the role of the computer as an information tool.

Bellanca, Joseph J. 1988. When all else fails, try communication. *American Industrial Hygiene AssociationJournal.* 49(11): This guest editorial is a brief sensible message about the art of communication in an industrial hygiene setting, and the importance of communicating effectively in order to accomplish your industrial hygiene objectives.

Celanese Corporation. 1983. Hazard communication rule. *Federal Register.* 48(228)Nov.25,1983:53282-53283.

Chemical Manufacturer's Association. 1983. Hazard communication rule. *Federal Register.* 48(228)Nov.25,1983:53283.

Covington, Michael A. 1985. Documentation that works. *PC Tech Journal.* 3(1):165–173.
Everyone complains about documentation, but few are as insightful as this linguistics professor. This article should be mandatory reading before writing documentation.

Date, C. J. 1986. *An Introduction to Database Systems.* Vol.1 (4th Ed). Reading, Mass.: Addison-Wesley Publishing Co., p. 487.
A classical work on relational databases, this is not for beginners, but it is a valuable reference for those interested in database theory and development.

Douglas, Paul. 1957. *Communication through Reports.* Englewood Cliffs, N.J.: Prentice-Hall, Inc. A timeless work on writing reports that itself is a perfect example of logic, clarity, and simplicity. The author believes in writing reports for a specific purpose: to be read.

Earnest, L. 1989. Can computers cope with human races? *Communications of the ACM.* 32(2):174–182.

This provocative article is long overdue. It is one man's experience in trying to fit people into a computer record that does not reflect the real nature of our society.

Ein-Dor, P., and E. Segev. 1978. *Managing Management Information Services.* Lexington, Mass.: D. C. Heath/Lexington Books.

Gould, John D., and Clayton Lewis. 1985. Designing for usability: key principles and what designers think. *Communications of the ACM.* 28(3):300–311.

User participation in design is the theme. A prototype is employed to find out how the users will use the system, not how the system matches the requirements analysis. Three principles are stressed: (1)understand the users, (2)use simulations and prototypes to measure the product, and (3) design, measure, and redesign iteratively. For both users and developers.

Harris, Thorne D., III. 1985. *The Legal Guide to Computer Software Protection.* Englewood Cliffs, N. J.: Prentice-Hall, Inc., p. 61.

Designed for computer users, programmers, software publishers, and business people, this book explains the legal reasoning and requirements of computer programming, software ownership, and software purchase. Written for laymen, it is a good desktop reference on copyrighted software and software contracts.

ILO Constitution. Washington, D. C.: International Labor Office.

ILO International Labor Convention No. 155. 1981. Occupational safety and health. Washington, D. C.: International Labor Office.
Washington,D.C.

ILO International Labor Convention No. 161. 1985. Occupational health services. Washington, D. C.: International Labor Office.

Industrial Safety and Hygiene News. Surveying the field: who's keeping bad records. September 1987:64–65.

Results of an *ISHN* poll of readers on the prevalence of inaccurate recordkeeping.

Kerr, Pamela S. 1978. Recording occupational health data for future analysis. *Journal of Occupational Medicine.* 20(3): 197–203.

An account of a Standard Oil Company integrated occupational health management system developed in-house for analyzing data on employee exposures and its impact on employee health. This was an early system of eight distantly located terminals connected to the Chicago office to discover the high risks of chemical, physical, and biological hazards.

Kleinman, Goldy D. 1984. Occupational health and safety: the Swedish model. *Journal of Occupational Medicine.* 26(12):901–905.

Sweden is a leading European country in the legislation of occupational health and safety. Their laws provide a framework for employers and employees to use when bargaining in plant safety committees. Swedish legislation emphasizes consensus and cooperation. We learn by studying health and safety practices elsewhere, and Sweden is a good place to begin.

Laumer, John, and Dale Shipman. 1987. EPA rules provide design basis for expert system. *Government Computer News.* September 25,1987:139.

An expert system developed by the National Safety Council in 1984 for microcomputer use to help people determine whether the Resource Conservation Recovery Act (RCRA) rules apply to them. Each screen systematically steps the user through waste generation questions, concluding with the user's status.

Lee, Jeffrey S., and William N. Rom. eds. 1982. *Legal and Ethical Dilemmas in Occupational Health.* Ann Arbor, Michigan: Ann Arbor Science Publishers.
An exceptional collection of papers covering ethics, science, and law as they relate to occupational health. All three fields must be applied to legislation because of the technical nature of occupational health. The ethics of occupational medicine are considered in relation to the marketplace concerns of industry, societal responsibility for the health of individuals, and the process of government regulation and deregulation. Food for thought for lawmakers at all levels of government and for occupational health professionals who need to work cooperatively.

Mabbett-Shoop, Barbara. 1987. OSHA recordkeeping: clarifying and defining issues and problem areas. *Industrial Safety and Hygiene News.* September, 1987:64–65.
This clarifies definitions for OSHA recordkeeping according to the U.S. Office of Management and Budget's interpretation of employee illness data and other data required for OSHA reports.

Markus, M. Lynne, and Niels Bjorn-Andersen. 1987. Power over users: its exercise by system professionals. *Communications of the Association for Computing Machinery.* 30(6):498–504.
A perceptive exploration of the relationship between information services and users that should be read over and over by systems people. The authors make clear the distinction between *having* power and *using* it, a critical difference that sharply affects the relationship between developer and user.

Moffett, J. D., and Morris S. Sloman. 1988. The source of authority for commercial access control. *Computer.* 21(2):59–69.

Navickas, Irene A. 1987. Training and the End User. Northbrook, Ill: Navickas and Associates.
A short paper published by the author for presentation at a user's group meeting. This is good advice from an experienced software trainer about the qualities needed for a trainer and the preparation needed to train.

O'Neill, Brian, and A. B. Kelley. 1985. Costs, benefits, effectiveness, and safety: setting the record straight. *Readings in Cost Benefit/Cost Control.* Park Ridge,Ill.: American Society of Safety Engineers.
Concepts of cost/benefit analysis and cost- effectiveness analysis are defined in relation to standards for motor vehicle safety. The concepts are compared in a simple, direct manner for the layman. A reprint of a 1974 address, but unfortunately, the ASSE garbled the first page. Begin reading on page 3 instead of page 2.

Pendergrass, John A. 1983. Interaction and communication with management. *American Industrial Hygiene Association Journal.* 44(5):313–315.
Written at a time when the author was a practicing hygienist in industry, this

paper is designed to help industrial hygienists provide information, advice, and assistance to line management. The author is a strong advocate of accurate, concise, understandable, and complete reports.

Perry, William E. 1987. Customers, not quotas, must come first. *Government Computer News*. November 20,1987:43–44.
Important lessons can be learned in an atmosphere of wry humor. This paper makes a significant point about the differences in viewpoint and objective between the data processing programmer and the user by comparing this to his experience with a car repair. Worthwhile reading for both developer and user.

Ross, Charles W. 1984. *Computer Systems for Occupational Safety and Health Management*. New York: Marcel Dekker, Inc.
The author provides excellent examples of safety systems that illustrate good use of coding. Most safety systems will not need as numerous codes as shown for the petroleum industry example, but there are a lot of ideas to be gained from reading this carefully.

Samuels, Sheldon F. 1977. The problems of industry-sponsored health programs. *Background Papers on Industry's Changing Role in Health Care Delivery*. ed. Richard H. Egdahl, pp. 152–158. New York: Springer-Verlag.
Explores the relationship between employees, the company physician, and the need for an integrated industrial-community program for early detection of preclinical disease in workers. Concerned about the conflict of interest of programs sponsored by industry and suggests that government needs to intiate control.

Tepper, Lloyd B. 1987. The strategic dimensions of occupational medicine. *Journal of Occupational Medicine*. 29(4):325–329.
Explores the place of occupational medicine in the corporate setting, examining the physician's responsibility to the organization and to the employee patients. Under the competitive conditions in which the company exists, the author advocates "an institutional commitment that promotes health, prevents disease, and improves the lives of people who work."

Weiner, Peter. 1980. Employee exposure and medical records. *Federal Register*. 45(102)May 23,1980:35220.

INDEX